Strategies and Technologies for Developing Online Computer Labs for Technology-Based Courses

Lee Chao
University of Houston-Victoria, USA

T0324981

IGI PUBLISHING

Hershey • New York

Acquisition Editor:	Kristin Klinger
Senior Managing Editor:	Jennifer Neidig
Managing Editor:	Sara Reed
Development Editor:	Kristin Roth
Copy Editor:	April Schmidt
Typesetter:	Michael Brehm
Cover Design:	Lisa Tosheff
Printed at:	Yurchak Printing Inc.

Published in the United States of America by
    IGI Publishing (an imprint of IGI Global)
    701 E. Chocolate Avenue
    Hershey PA 17033
    Tel: 717-533-8845
    Fax: 717-533-8661
    E-mail: cust@igi-global.com
    Web site: http://www.igi-global.com

and in the United Kingdom by
    IGI Publishing (an imprint of IGI Global)
    3 Henrietta Street
    Covent Garden
    London WC2E 8LU
    Tel: 44 20 7240 0856
    Fax: 44 20 7379 0609
    Web site: http://www.eurospanonline.com

Library of Congress Cataloging-in-Publication Data

Chao, Lee, 1951-
  Strategies and technologies for developing online computer labs for technology-based courses / Lee Chao, author.
    p. cm.
  Summary: "This book discusses design strategies, implementation difficulties, and the effectiveness of online labs. It provides scholars, researchers, and practitioners support for lab-based e-learning, gives guidance on the selection of technologies for various projects, and illustrates Web-based teaching with case studies"--Provided by publisher.
  Includes bibliographical references and index.
  ISBN 978-1-59904-507-8 (hardcover) -- ISBN 978-1-59904-509-2 (ebook)
  1. Computer networks--Design and construction. 2. Instructional materials centers--Design and construction. 3. Schools--Computer networks--Design and construction. 4. Educational technology. 5. Web-based instruction. I. Title.
  TK5105.5.C4594 2008
  607.8'5--dc22
                    2007022232

British Cataloguing in Publication Data
A Cataloguing in Publication record for this book is available from the British Library.

Strategies and Technologies for Developing Online Computer Labs for Technology-Based Courses

Table of Contents

Section II:
Design of Online Computer Labs

Section III:
Development of Online Computer Labs

Section IV:
Management of Online Computer Labs

Section V:
Trends and Advances

Foreword

Creation of online computer labs for use in computer science and information systems courses is a matter that is widely discussed in professional meetings and at other formal and informal meetings of computing faculty. Mostly the conversations would revolve around statements of wishful thinking, difficulties encountered, and partial successes. Occasionally, one meets someone with a success story, sharing his or her experiences to an eager group of individuals. The value of Li Chao's *Strategies and Technologies for Developing Online Computer Labs for Technology-Based Courses* is greatly enhanced by the fact that it is perhaps the first book to appear on the difficult task of planning and developing an online computer lab in support of IT courses. The book has importance that is not only timely but timeless, in that it discusses many of the underlying issues besides the latest systems and technologies. To its author, who I have had the pleasure of working with for many years, may every reader render due appreciation, for distilling the wisdom garnered from his experiences.

May I underscore that this book is not just theory—even though the author has gone to extraordinary measures to describe the methodological underpinnings—but rather based on actual experience, what we actually use, the result of many trials and tribulations and the ultimate lessons learned. What Chao describes here in these pages actually works, and in many instances reflects an optimal strategy for the set of conditions being considered.

I was glad to write the foreword to this new, innovative, and highly relevant book, written by one of our most popular teachers.

Meledath Damodaran, PhD
Professor and Coordinator
Computer Science, Information Systems and Math
University of Houston-Victoria
Victoria, Texas

Meledath Damodaran is a professor of computer science and mathematics at the University of Houston-Victoria. He is the coordinator of the Computer Science and Mathematics Programs at UHV. Prior to joining UHV in August 1991, he taught computer science at various universities for over 11 years.

His research interests are in software quality management, parallel processing, and neural networks. He is also interested in software engineering, software project management, and computer science and computer information systems education.

He teaches a wide variety of courses including operating systems, software project management, software engineering, programming Language theory, computer security, database design, artificial intelligence, computer architecture, and information systems.

Dr. Damodaran has also worked as a consultant to industries and governmental agencies. He was a Fulbright scholar (1992) and a Mellon visiting faculty fellow in the Computer Science Department at Yale University in New Haven (1989 to 1990). Dr. Damodaran has also served as the President of Applied Computer Consultants Inc., in Edmond, OK (1984 to 1986). Dr. Damodaran received his doctorate from Purdue University (1977).

Preface

Today's e-commerce environment requires our computer science and information systems students to have hands-on skills in developing and managing information systems to support the Internet based business of any size. It requires our students to be able to handle the client-side jobs such as application development and other programming related tasks as well as server-side operations such as network and system administration. To meet the requirements of the IT industry, our students must have hands-on practice on software and hardware used in an IT infrastructure. Therefore, various computer labs are needed to support the hands-on practice in various technology-based courses.

On each campus, large or small, computer labs are the key component for a computer science or computer information systems department. The construction of computer labs involves various technologies. A computer lab consists of computers, operating systems, application software, system and network management software, security management software, remote access software, network equipment, storage devices, and remote access gears. These technologies are used to provide services, such as supporting a computing environment for students to perform hands-on practice, allowing faculty members to demonstrate the use of new technologies and to develop new course materials, supporting various online courses, storing teaching materials, and providing remote access service for students to remotely log on to the computer labs on campus.

More and more courses are taught online now. Many academic fields have developed effective solutions for online teaching and learning. Generally, for classes with no lab activities such as English, history, and education, the implementation of online courses are relatively easy. Many commercial software packages such as WebCT are available to support course setup for these types of courses.

Given the fact that online teaching has become the mainstream in higher education, one may assume that most of the technology-based courses that require hands-on practice have already been taught online now. However, not many papers and reports are about online teaching in this area, especially about courses that require server-side hands-on practice. For many information systems courses, WebCT or similar software packages are not sufficient because these software packages do not support hands-on practice on IT products which is a very important part of the information systems curriculum. Unlike a WebCT-based online course which mainly processes online course content through client-side application software, for a technology related curriculum, the online courses need to develop and maintain server-side projects through the Internet, such as managing operating systems, building client-server computing architecture, developing enterprise-level database servers, and providing e-mail services. An online technology-based course needs an online computer teaching lab whose implementation is a challenging task.

The IT industry is a rapidly developing industry. Each year, even each month, new technologies are created. To teach students the knowledge that is not obsolete, computer labs need to be updated according to trends. Faculty members have to keep adding new course content to the materials for teaching and hands-on practice and to request that the computer labs be updated accordingly.

People who are specialized in various technology fields are involved in the lab development and management. Employees of technology consulting companies, managers and technicians in computer service departments, administrators at different levels of a university system, computer science or information systems faculty members and students are all involved in the construction, development, and management of online computer labs. The development of online computer labs requires various skills and specialties such as management skills of administrators and IT managers; it needs technicians with specialties in networking, system administration, database administration, application development, and security management. It often requires faculty members to participate in the construction and management of an online computer lab since the online computer lab is used to support classes taught by faculty members. Students are the ultimate users of computer labs. More importantly, the development of online computer labs needs support from the university's top administrators. People who are involved in the online computer lab development process need to have a strategic view about the computer lab. It is helpful for these people to understand the theory and practice of Web-based teaching while developing the online computer lab. They need to know the technologies involved in the process, and they need to know how to efficiently manage the online computer lab and know the current technology trend.

The development and management of online computer labs needs financial support. They usually take a large portion of a university's budget. The cost of software, hardware, lab maintenance, and labor can be significantly high. We need an annual budget to cover these costs. Therefore, developing online computer labs requires knowledge in financial planning, expenditure monitoring, equipment purchasing, and on how to allocate resources for the labs. We need an accountant from the university's finance department to keep track of the costs and to make sure that all the spending follows the regulations set by the government and the university.

As mentioned above, an online computer lab itself is a complex project. It usually requires a great deal of effort to construct, develop, and manage an online computer lab. It is crucial to know the strategies in developing online computer labs. This book is designed to address these issues. It will help the reader, either a veteran administrator or someone who has just started his/her IT career, to get a deeper understanding of developing online computer labs and to be able to handle enterprise-level computer lab development tasks.

This book investigates technology-based courses that require online computer labs to support teaching and hands-on practice. It discusses the difficulties of online computer lab implementation and management. It also talks about the online computer lab planning and budget related issues. The book presents different approaches to accomplish the lab construction tasks. It provides information about different types of online computer labs and the technologies used in these computer labs. It gives comparisons on the advantages and disadvantages of each type of lab.

This book provides related information and technologies to help faculty members develop teaching materials that require online hands-on practice. The book shows instructors how to create online lab activities in today's Web-based teaching environment. It also provides information to help technicians and faculty members manage online computer labs, especially, security management and lab update strategies.

For teaching and learning with online computer labs, this book discusses various methods and tools for lab testing and evaluation. For future development of online computer labs, the book gives information about the trends in Web-based teaching and technologies that can potentially improve Web-based teaching and learning.

The Challenges

For technology-based courses taught online, hands-on practice relies on online computer labs. However, many technology-based courses are still not fully supported by online computer labs. There are some reasons for online computer labs not being able to fully support online technology-based courses. One reason could be lack of funding to purchase necessary equipment for developing a remotely accessible computer lab. For many small campuses, funding for computer labs is very limited. Online computer labs require reliable remote access service and network equipment to ensure better performance. This necessary equipment costs money. An online computer lab also requires a stable electricity power supply. Additional Uninterruptable Power Supply (UPS) equipment is necessary to keep the servers running.

Lack of experience is another reason for online labs not being able to fully support the online technology-based courses. Web-based teaching has only a short history. Universities have put their main focus on those courses supported by commercially available learning management software packages such as WebCT. It is relatively easy to use WebCT or similar software to support courses that do not require hands-on practice on commercial IT products. On the other hand, for online technology-based courses, we need to do more than just use the commercial learning management software. Putting a local campus computer lab online for technology-based courses is relatively new to many computer service departments. It needs experienced technicians to implement remote access service and enforce security measures to keep the internal network from attacks of Internet hackers. The fast changing technology and course content require the technicians to keep up with the technology trends. Online computer labs are also new to faculty members and students who need training on using technologies involved in online computer labs.

Another reason is that it is difficult to manage an online computer lab. When many students log on to the same computer lab, it significantly slows down the lab performance. The technical support team needs to resolve the bottleneck problem. This team is also expected to help students solve technical problems on the client side. Unlike the help in a computer lab which has the same type of computers, helping online students is a much tougher job since the computer systems on the students' side are different in brand and model. The students' computers may have different operating systems installed and may be configured differently. Also, it is difficult to provide technical support remotely since the students are not familiar with the technologies used in the online computer lab and the technical support team may not know the course content well enough to provide helpful advice without face-to-face conversations with the students.

Among the reasons of not being able support a fully online computer lab, lack of support from university administrators is the one that has the greatest impact. It will directly influence the decisions on the computer lab budget and on providing experienced personnel to run the computer lab. It will also influence the decisions on computer lab room maintenance, electricity power supply, class scheduling, release of workload for faculty members to develop the online computer lab, and many other services.

As mentioned above, it is a challenging task to develop an online computer teaching lab for technology-based courses. To face the challenge, extra effort is needed to overcome the difficulties. With the right strategies and tools, most of the difficulties can be overcome.

The Answers

To develop a successful online computer lab to support teaching and hands-on practice for technology-based courses, the lab designer/developer and manager need to understand the requirements for the teaching and hands-on practice and the overall structure of a Web-based teaching system. The following are some key areas that need to be dealt with during an online computer lab development process.

It is important to know what the requirements are for the online computer lab. The lab should be designed to meet these requirements. Therefore, the first step in the lab developing process is to understand the Web-based teaching environment and practice. Understanding the curriculum of computer science and computer information systems can also help the lab designer/developer and manager make the right decisions in developing the online computer lab. To better understand the use of the online computer lab, the people who are involved in the lab development need to collect information about how students perform their lab activities in the lab.

The lab designer/developer and manager need to collect information about constraints that may impact the design decisions. They need to investigate the limitations on the budget, technical support, size of the computer lab room, and even the power supplies. Before the design process can get started, it is necessary to review the university regulations and security policies.

A design process needs to select the technologies that can be used to achieve the design objectives. However, the selection of technologies is limited by the constraints and university policies. Therefore, careful planning and budgeting are required before the construction of the online computer lab starts. The design should be based on the factors such as maximum number of possible concurrent users, the number of courses that will share the lab resources, the requirements for the computer and network equipment, the requirements for security, and the requirements for performance. Often, a design process is not linear. It requires several rounds of testing and modifications before the final version of the design can be completed.

Once the equipment is purchased, the implementation process can be started. The physical implementation may encounter various technical problems. It requires troubleshooting skills and experience. The implementation process should be closely monitored so that the project does not get behind schedule. It needs strong support from the computer service department and/or software and hardware vendors. Universities may also outsource the lab implementation tasks to a consulting company. In such a case, the collaboration with the consulting company is crucial for the success of the implementation.

When an online computer lab is in use, we must provide helpful technical support and training for lab users. Detailed lab usage instruction should be available to all students and faculty members. A knowledgeable technical support team should be formed to help faculty members and students solve their technical problems. Faculty members may also need a technicians' help when developing hands-on practice materials.

A well managed online computer lab is the key for success. Security is always the top concern. The lab manager needs to be aware of various security vulnerabilities and know how to enforce security measures to protect the internal network. The lab manager should be familiar with the daily maintenance procedures and know how to deal with emergency problems. Usually, the lab manager and faculty members need to work together to redesign and reconstruct the lab when new course content is added to online courses.

A well-designed online computer lab alone is not enough for successful Web-based teaching. We need the teaching materials to be suitable for Web-based teaching. Due to the fact that there is no face-to-face communication in online courses, the teaching materials for hands-on practice need to be carefully designed to provide as much detail as possible so that ambiguity can be reduced to a minimum. Instructors need to know how multimedia materials can be used to improve online teaching and what the limitations are. They should also know how to use the technologies to develop multimedia course content.

For further improvement, we need the feedback from instructors and students. Lab testing and evaluation are also important for developing and managing a successful online computer lab. The lab manager should be familiar with the procedures of lab testing and evaluation. He/she needs to know how to design an evaluation instrument and the procedure to collect information.

This book will address each of the key areas in online computer lab development. It will give an overview of hands-on practice in a Web-based teaching environment and provide background information about the above key areas. In this book, we will discuss the design strategies, implementation methods, and the effectiveness of online computer labs. The book will present some possible solutions for the challenges raised in an online computer lab development process. It will provide readers with detailed information about the technologies involved in lab construction and management.

Organization of the Book

This book is organized into five main sections. The following is a brief description of each section.

Section I: Introduction

This is an introduction section which includes two chapters.

The first chapter provides background information about Web-based/online teaching and technology-based courses. This chapter discusses the strengths and weaknesses of Web-based teaching systems. Then, it analyzes some technology-based courses and their requirements for teaching and hands-on practice. This chapter also investigates how the strengths and weaknesses of Web-based teaching will affect the teaching and learning in technology-based courses.

The second chapter provides information about technologies used in the development of online computer labs. It describes the functionalities of these technologies and the roles played by the technologies.

Section II: Design of Online Computer Labs

This section discusses the issues related to the design of online computer labs. It includes Chapters III and IV.

Chapter III presents a systematic way to identify the requirements for an online computer lab. It then discusses the issues related to project planning such as budgeting, scheduling, and organizing.

Chapter IV discusses the strategies for developing online computer labs. The content in Chapter IV is used to prepare lab designers/developers to transfer the design into the technical stage. It covers the modeling process, physical design, and resource sharing issues. The issues on the selection of hardware and software, network equipment, and remote access technologies are also covered in this chapter.

Section III: Development of Online Computer Labs

This section covers the issues related to the development and implementation of online computer labs. Chapters V through IX are included in this section.

Chapter V discusses the issues involved in the implementation of servers, used to support the daily operations of an online computer lab. Various server implementation and server deployment plans are presented in this chapter. This chapter also deals with the server-side configuration issues.

Chapter VI is about the network development for an online computer lab. This chapter deals with the issues related to network configurations to meet the teaching and hands-on practice requirements. It discusses the strategies on how to work with a network that consists of different technologies. This chapter talks about the issues related to the implementation of networks for different types of online computer labs. It also provides information about various network equipment.

Chapter VII focuses on the client-side computing environment. It examines the issues related to the installation and configuration of students' home computers, software, and network devices so that this equipment can be used to remotely access the servers installed in an online computer lab. This chapter provides configuration related information on remote access technologies. It also provides information about the configuration of multimedia devices.

Chapter VIII covers the topics related to security. An online computer lab is exposed to various malicious viruses. It is important to establish security policies for an online computer lab. Therefore, this chapter discusses security policy issues. It also investigates the sources of security vulnerabilities. It provides some strategies in dealing with the vulnerabilities. The technologies used to enforce the security policies are also covered in this chapter. Since some of the technology-based courses require students to have an administrator's privilege in order to carry out their hands-on practice, this may create a serious security problem. This chapter discusses some possible solutions to resolve the conflict. To educate lab users about security vulnerability prevention, this chapter also presents ideas about how to deliver security instructions to the lab users.

Chapter IX deals with the issues related to the development of online teaching materials. This chapter first looks into the design of lab-based teaching materials to meet the requirements of hands-on practice. The teaching materials for hands-on practice are categorized into three types, text-based, combination of text and figures, or multimedia based materials. This chapter provides strategies on creating and using these three types of teaching materials. It also investigates the limitation of each type of teaching materials. The development of multimedia teaching materials requires additional tools. This chapter demonstrates several approaches to develop multimedia teaching materials. It also includes information about the tools used to develop video and audio course content.

Section IV: Management of Online Computer Labs

This section focuses on the lab management related issues. Chapters X and XI are included in this section.

Chapter X discusses the special maintenance needs for online computer labs. At the beginning of this chapter, it discusses the computer lab policy to be enforced in a computer lab. It also discusses the issues related to the technical support and the daily maintenance. It presents the strategies for system backup, recovery, and performance tuning. Online computer labs often need to be rebuilt after each semester. This chapter provides some solution for accomplishing this task.

Chapter XI is about testing and evaluating an existing computer lab. It begins with the discussion of requirements for testing and evaluation. It then investigates various ways to carry out the testing and evaluation process. An evaluation process can be used to collect feedback from the lab users. This chapter introduces some lab evaluation instruments which can get the job done. The last topic covered by this chapter is about the measurement of effectiveness for using an online computer lab.

Section V: Trends and Advances

Chapter XII is about the trend in Web-based teaching that has the impact on the development of computer labs. This chapter provides information about some of the trends in Web-based teaching and the trend in technologies. These trends reflect the changes in e-learning structures, management, content development, and the software and hardware used in the development of computer labs.

Acknowledgment

My thanks go to the book reviewers who have provided comprehensive and critical comments which are valuable contributions to the book. I am grateful to Dr. Mel Damodaran who has written the wonderful foreword for this book. I would like to give my deep appreciation to Dr. Jenny Huang for her review on the content of the manuscript and many constructive suggestions to enhance the quality of the book. I would like to acknowledge the help of my colleagues and the university's technical support team with the online computer labs. Special thanks go to all the editors and staff at IGI Global Inc. especially to Jan Travers, Meg Stocking, Jessica Thompson, Kristin Klinger, Jennifer Neidig, Sara Reed, Kristin Roth, April Schmidt, Michael Brehm, Lisa Tosheff, and Mehdi Khosrow-Pour. They have provided great help from the decision on the book title to the final publication of this book.

Section I

Introduction

Chapter I

Introduction to Online Teaching of Technology-Based Courses

Introduction

With the improvement of the Internet and computer technologies, online or Web-based teaching has become an important teaching and learning method in educational institutions. In various degrees, online teaching has been implemented in almost every higher education institution. To better understand online teaching systems and how they are related to the book's main topics, online computer labs for technology-based courses, we will overview online teaching and technology-based courses in this chapter. We will take a look at the strengths and weaknesses of the Web-based teaching (WBT) systems. We will also investigate the roles played by these Web-based teaching systems in teaching technology-based courses. The investigation of these aspects will lead to the discussion to the book's main topics.

Although Web-based teaching has a short history, it is one of the fastest growing areas in higher education. The rapid development of Web-based teaching certainly makes it difficult to include everything in one chapter. Therefore, we will first have a very brief overview about Web-based teaching. After the brief overview, we will consider the advantages and disadvantages of Web-based learning systems. For the disadvantages, some suggestions will be given to reduce the impact of the disadvantages. Then, we will briefly discuss technology-based courses and their requirements for hands-on practice. Next, we will investigate the challenges on implementing Web-based teaching for technology-based courses. The last paragraph of this chapter will summarize what we have discussed in this chapter.

Background

Online education has attracted many researchers' attention to the issues related to computer assisted teaching and learning. Many research studies have been done to study the impact of technology on teaching and learning. In the debate on the impact of technology, Kozma (1994) pointed out that the media and methods together would impact teaching and learning. In this chapter, a research approach was developed to analyze how learners interact with instructional designs and how the media and methods influence understanding. Nott, Riddle, and Pearce (1995) studied the impact of WWW technology on learning. Mixed reaction towards the role of WWW technology was reported in their study. Chu's (1999) article analyzes the development of a Web-based remote interactive teaching system and addresses the issues such as assessment, cost, students' feedback, and collaboration. It also discusses the positive and negative sides of remote interactive teaching systems. In this chapter, we will further analyze the advantages and disadvantages of e-learning. Since technology-based courses and Web-based teaching all rely heavily on technology, there is no doubt that a well-developed instructional design will greatly influence teaching and learning.

The use of computers and the Web as a teaching and learning tool was gaining momentum in the 1990s. The survey conducted by the Sloan Consortium (2005) presents the following results: Among all the graduate schools offering face-to-face courses, 65% of them are also offering online graduate courses; among all the undergraduate schools offering face-to-face courses, 63% of them are also offering online undergraduate courses; among all the Master's degree programs offering face-to-face courses, 44% of them are also offering online programs; among all the business programs offering face-to-face courses, 43% of them are also offering online programs. Based on these figures, the Sloan Consortium (2005) concludes that online education has entered the mainstream.

Computer-based training (CBT) is the training delivered via computers. However, CBT does not necessarily deliver the training through the Internet. On the other hand, Web-based training (WBT) is the computer-based training delivered over the

Internet and accessible using a Web browser. Numerous articles and books about e-learning, CBT and WBT have been published. The use of computers to assist instruction started long before the creation of the Internet. Hall (1970) reported the use of computer-assisted instruction in Pennsylvania around the late 1960s. The potential of World Wide Web to be used to carry out computer-based training was mentioned in the early 1990s (Szabo & Montgomery, 1992). Since then, a large number of books and research articles about WBT have been published. In the late 1990s, more detailed studies about the essential tools and techniques for planning, creating, and implementing WBT were published. For example, Hall's (1997) book illustrates the process of developing WBT courses. It includes many well-designed WBT Web sites available at that time to illustrate the WBT design process.

At the beginning of this century, books about hands-on and practical guidance for designing a successful WBT were available for readers. Horton's (2000) book gives a good overview about WBT. More importantly, his book gives instructions on how to develop multimedia courses with courseware products; those products are software packages used to create Web pages for online teaching.

Recently, more studies in the WBT design area have been published. Readers can find the history and experiences in designing WBT for e-learning summarized by Driscoll (2002), whose book provides readers with WBT designs from the pedagogical point of view. The knowledge is helpful for many WBT designers who are usually technically-oriented instead of instructionally-oriented. Experienced WBT designers who are interested in investigating WBT design and implementation strategies at the top level can refer to the book by Driscoll and Carliner (2005).

As technology advances, many online courses have included multimedia-based teaching materials which can greatly improve the quality of Web-based teaching. To learn how to design multimedia-based instruction for CBT and WBT, readers can refer to the book by Lee and Owens (2004). Their book covered instructional design methodologies and some well-defined planning strategies.

A growing number of higher education institutions utilize the Internet to reach out to more students, to reduce the cost and to improve the quality of teaching. As more and more courses are taught online now, many of these higher education institutions have developed effective solutions for online teaching and learning. For courses in the information systems or computer science related curricula that do not require lab activities, the implementation of online courses is relatively easy. Many commercial learning management system (LMS) software packages such as WebCT are available to support course setup, management, and assessment. Usually, these LMS software packages also support multimedia functionalities and collaboration utilities such as chat rooms and discussion groups. On the other hand, many courses in computer science and computer information systems curricula require hands-on practice on IT products. To allow hands-on practice using a computer lab over the Internet, especially for the server-side hands-on practice, the implementation of Web-based hands-on practice will require additional effort. It is necessary to support this type of Web-based courses with fully functioning online computer labs.

Developing an online computer lab is a complex process. A lab development project needs to be thoroughly planned. It involves various technologies. It must be carefully designed before it can be implemented. After an online computer lab is constructed, you need a group of experienced computer service personnel to manage the lab and provide the necessary training for lab users. To begin with, it is necessary for a lab designer to understand the Web-based teaching environment and the requirements by various technology-based courses. In this chapter, we will discuss these issues.

Web-Based Teaching

By the end of the 1990s, the number of the Web sites had grown over 5,000,000. These Web sites provided a large amount of information stored in text, graphic, video, and audio formats. The hyperlinks created on the Web pages allowed readers to navigate from one Web page to another to get related information. With such a convenient communication tool, companies and universities began to post information on their Web sites.

Use of Web Sites

Information posted on a university's Web site may include class schedule, course catalog, handbook, office information, and class handouts which may contain instructions such as how to run Java codes on a personal computer, and so on. By posting class handouts on the Web sites and making them available to students, universities began to use the Web as a teaching and learning tool. In the teaching and learning process, the Web can be used to:

- **Post teaching materials:** Web sites can be used to post lecture notes, lab manuals, syllabi, assignments, research reports, and many other things.
- **Manage online courses:** On the Web, instructors can grade the assignments and record student grades.
- **Support collaboration activities:** Through the Web, instructors and students can communicate with each other through e-mail, chat rooms, discussion groups, virtual whiteboards, or online conferencing.
- **Perform hands-on practice:** As the Internet technology advances, students can now access an online computer lab for hands-on practice.

Web-Based Teaching

Teaching or learning through the Web is called Web-based teaching (WBT) or Web-based learning (WBL). A WBT system provides distance learning with multimedia content and allows interactivities through the Internet. Similarly, the Web is also used by companies to carry out employee training. Training delivered through the Web is called Web-based training (WBT). There are several similar names for Web-based teaching systems. For example, the Web-based instruction system (WBIS), Web-based teaching and learning system (WBTLS), and e-learning system are also often used in various academic fields. In this book, we will use the term e-learning for general technology-based teaching and learning and WBT for Web-based teaching and learning. WBT can be considered as Web-based training or Web-based teaching.

Improvements of Web-Based Teaching

After a decade of development, today's WBT is no longer just posting class handouts. Now, WBT systems are becoming a strategic component of higher education institutions for teaching, learning, and class management. The Sloan Consortium (2005) report indicates that 53.6% of the academic institutions they surveyed agree that online education is critical to their long-term strategy. Most of these institutions conveyed that the quality of online instruction is equal to or even better than the quality of traditional instruction.

WBT is now a billion dollar business and is supported by hundreds of learning management systems. A learning management system (LMS) is used to host and manage course Web sites which post online course materials. Through LMS, instructors can manage classes, grade assignments, and announce updated information to students. LMS is also used by administration offices to handle the enrollment and tuition payment. LMS is now one of the leading tools that most higher education institutions use to build, deploy, and support WBT.

As the Internet technology advances, the interactivity, as one of the major concerns about e-learning, has been improved significantly. Technologies such as chat rooms, discussion groups, blogging, VoIP, and virtual classrooms allow students to interact with instructors and to work together as a team. We are expecting even better collaborating tools to be available in the next few years.

Today's teaching and learning environment in higher education encourages further development of e-learning. More and more students grow up with computer technology. These students have adequate computer skills to participate in WBT. Faculty members have also made progress in mastering computer technology. They feel more comfortable using technology to develop course materials and to manage classes through the Internet. With more experience in setting up e-learning systems, the technology support team can now provide better and faster services. More and more powerful and convenient course development and management tools are available and continue to improve. According to the report by Oblinger and Hawkins (2005), well over 100,000 distance education courses are offered today. As they point out, online enrollment is increasing not only in number but also in growth rate. The growth rate was 19.8% in 2003 and the growth rate for 2004 was estimated as 24.8%.

The strength of WBT makes it become one of the major teaching platforms that provide higher education to millions of students. However, like many others things, WBT has its weaknesses. WBT can be greatly improved by overcoming these weaknesses. In the next topic, we will have an overview of the strengths and weaknesses of online teaching. The understanding of the strengths and weaknesses of WBT will lead to a better design of an online computer lab.

Strengths and Weaknesses of Web-Based Teaching

Each of the e-learning systems has its own strengths and weaknesses. In some circumstances, it is not always the best way to teach a technology-based course through an e-learning system. Sometimes, we have to use e-learning not because it is the best in all aspects. It is because some of its strengths we will not be able to get from other teaching/learning systems. Often, we gain something from e-learning systems and lose something that may be the strength of other systems. As pointed out in the Background section, e-learning has entered the mainstream of higher education. It must have some strong points that other methods of teaching/learning can not match. In the following, let us investigate the strengths of e-learning systems.

Far-Reaching Availability

Availability is one of the most important strengths of e-learning. Once an e-learning system is created, it can be accessed globally. Students are able to access course materials from anywhere and at anytime. Traveling to campus is not required. For colleges and universities that are located in rural areas, their campuses are usually small and spread out in a large area. It often requires that students travel dozens of miles or even over a hundred miles to take classes. In some remote areas, some of the students may spend several hours on the road. It is difficult for the faculty members, too. They need to travel from campus to campus to teach classes. Time on commuting could be better spent on helping students or doing research. At colleges and universities located in the inner city, students and faculty members have their commuting problems, too. A traffic jam on the highway often causes unexpected delay.

In today's colleges and universities, a large number of nontraditional students enrolled. While taking courses to pursue their academic degrees, these students are have responsibilities for their jobs, families, and communities. It is often difficult for these students to find the right time that fits in the class schedule. Especially on those small campuses, only limited numbers of classes are offered in each semester. Also, some jobs require these nontraditional students to travel to other places. This makes the difficult situation even worse. After missing some classes, they may be behind in their homework assignments. Not only do they have to catch up with the course material they have missed, they also need to learn the new course content to keep up with other students. As the frustration builds up, the situation often leads to dropping out of the classes. The high availability of e-learning makes it possible for those students who cannot take certain classes at a specific time to take these classes at the time that is the most convenient for them. Students are able to study course materials, perform hands-on practice, and complete assignments anywhere across the world. For these students, e-learning is necessary. There is no other teaching/learning platform that can provide the same level of availability.

Greater Flexibility

Flexibility is also one of the strengths of the e-learning platform. E-learning is far more flexible than the traditional classroom teaching. No classroom is necessary for the e-learning platform. Students can study the course material in public libraries, in their homes, at airports, and wherever there is an Internet connection. From anywhere, students can submit their homework and take quizzes. Or even, if permitted, take exams from remote locations. Students who have a busy working schedule can use weekends and holidays to catch up with their class work. Once the multimedia course material is posted on a class's Web site, students can watch the lectures repeatedly if necessary.

For the traditional classroom teaching platform, a lecture can only be given once by an instructor in a semester. If students have difficulty understanding some parts of the lecture, they will not be able to go over the lecture again. This is especially hard for some foreign students who have limited listening comprehension in English. With the e-learning platform, these students can go through the same lecture as many times as they need. They can focus on the difficult concepts and theories and go over the instructor's explanation again and again. In that sense, e-learning does improve students' learning. Instructors can also benefit from e-learning. They can give lectures in the evenings, at weekends, and at anytime that is convenient for them. Wherever computer equipment and Internet connections are available, instructors can prepare their lectures and upload course materials to their course

Web sites. With the e-learning platform, instructors can also dynamically modify the course content based on their students' responses. They can correct misspellings, add more content, and emphasize certain content that is often misunderstood by students.

Cost Efficiency

Overall, e-learning costs less than the traditional classroom teaching platform. At the beginning, the e-learning platform may cost more. For example, a technical support team needs to be in place and the Internet connection needs to be established. In fact, these things cost a lot of money. Developing online teaching materials will also cost some money. Once the e-learning platform is established, the cost for e-learning operations is often less expensive. First, the saving on traveling is significant. Since classrooms are no longer necessary, there is no need to build new buildings even if there is a large increase in student population.

There is also a big saving on classroom facilities. For any sized colleges and universities, the cost on physical computer labs is substantial. With the e-learning platform, physical labs can be replaced by virtual labs and simulations. The cost for updating several servers that host the virtual labs is much less than updating several computer labs. The cost of maintaining a virtual lab is also significantly less than that of maintaining a physical lab. The overall expense on a physical lab is more than that on a virtual lab. The spending on a virtual lab often means adding more RAM and hard drives to existing servers, which costs much less than adding physical computers, printers, lab desks, and chairs to an existing computer lab. E-learning can also save on the costs of printing, electricity, parking, libraries, and many other things. In the long run, e-learning is often an economical choice.

More Maintainability

With the assistance of e-learning management utilities, the grading process can be greatly benefited from the e-learning platform. Instructors can grade students' homework assignments anywhere and at anytime. As soon as the homework is submitted by the students through the Internet, the instructors can grade them online. They do not have to wait to collect the homework in the classroom. It is easy to check if a student has submitted his/her home assignment for grading. It also saves a lot of time for the instructors to grade the assignments. They do not have to reorganize the submitted assignments into different folders, record the students' scores, and calculate the semester average. These tasks can all be done by the learning management service utility included in an e-learning system. As soon as the assignments are graded, the grades will be posted on the course Web site. The students can track their assignment submission records and scores through the Internet. The e-mail, discussion groups, and announcement tools provided by an e-learning system can assist communication among the students and instructors. These tools allow the students to get help from their classmates and instructors. The instructors can also use these tools to announce the changes of the course content and of the schedule to everyone in the same class.

Learning Enhancement

Through e-learning, students learn not just the content of the course, but they also learn how to use technology. The hands-on skills of using technology give students the competitive edge in their job hunting. With the experience of using technology, students can be more successful at work. E-learning encourages students to be more independent and more self-disciplined. When learning new concepts and theories, each student has his/her own learning style. They can choose the learning method that is most suitable for them. They can choose to watch the video, listen to the audio, or simply read the lecture notes.

After we have reviewed some of the advantages of e-learning, let us examine some of the disadvantages of the e-learning platform. E-learning is not always better than the traditional classroom learning. Being aware of the weaknesses of e-learning will help us avoid these weaknesses as much as possible during the planning, designing, and implementing phases of an online computer lab development process.

Lack of Face-to-Face Communication

The top advantage of traditional classroom teaching is the face-to-face communication. For example, when an inexperienced student is trying to figure out the configuration error of a lab-based project, nothing is better than a live instructor or a group of classmates who are working with the student and can help him/her on the spot. Often, a debugging process is a trial and error process. Without looking at the computing environment and the displays on the student's home computer, it is difficult for the instructor to guess what is wrong. Remember that most of the students who are taking a class may know very little about the course content of the class. Therefore, it is difficult for them to use professional terms to describe their computer problems happening on their side. A student may make a mistake in a configuration file; the error may appear in the next system boot-up process. It is hard for the inexperienced student to relate such an error back to the configuration completed before. When the student sends the information and question to the instructor for help, due to the lack of sufficient information, the instructor may not able to quickly pinpoint the problem. It may take some time to get the whole picture of the problem. The same situation often happens with the online technical support as well.

Student's frustration towards the e-learning platform can be shown in their class evaluations. Thus, it will not be surprising if face-to-face classes receive better student evaluations than online classes. It is much easier for a student to ask questions and for an instructor to give explanations if both the student and instructor are working in front of the same computer screen. Class discussions and group activities are also the advantages of traditional classroom teaching. When creating a Web-based course, it is crucial to resolve these issues, especially for hands-on activities. In later chapters, we will discuss some of the solutions to overcome the difficulties of online teaching and learning. It is risky if the teaching of an online technology-based course is unsuccessful. It may result in higher drop out rate and bad class evaluations, and damage the reputation of an educational institution.

More Dependent on Technology

E-learning requires highly reliable servers. The servers need to be powerful enough to provide adequate performance to support teaching/learning activities and to have enough storage space for the data generated by all the online courses. The network that connects the servers and students' computers should have enough bandwidth for Internet transactions.

Security is another major concern for the e-learning platform. Unlike e-learning courses in other academic fields where students and instructors work as clients of the server, in a technology-based course, both the students and instructors must have the server administrator's privileges. The students need to be the administrators to learn about server development and management. They need hands-on experience with networking equipment and operating systems, along with e-learning tools. Letting students have the administrator's privileges can certainly make the network administrator nervous.

The regular e-learning system only allows the instructor and students to use the application software to create teaching materials, browse the course Web site, download and upload course materials, and manage the course. Heavy security prevents students from performing any system administration tasks. For technology-based courses, students need to work on projects such as network configuration, system management, database administration, Web-based application development, and software and hardware maintenance. Students have to learn these skills to meet the requirements from the e-commerce industry. There is great resistance from college and university technical support teams who worry about the network security. Many of the technology-based courses have a tough time getting the class started due to the security constraints. The heavy security measures prevent students from logging on, through the firewall which is protecting an institution's network.

In addition to the security concerns, there are some other technical issues. For example, power outage and network problems often cause the server to be unavailable to students. On the client side, students need to use their own computers, network devices, and the broadband Internet connection to handle e-learning. Some students' equipment may not meet the technical requirements. Students may also have difficulty configuring the client side software for connecting to the server. All these technical problems can cause learning stress. Therefore, a knowledgeable and motivated technical support team is crucial for the success of e-learning. However, some technical service departments may not be able to fully cooperate. Later, we will discuss some strategies of forming a support team.

Increase of Work Load for Instructors

Developing a successful Web-based instruction system needs extra time to design, implement, and test the teaching materials. The process of producing video, audio, and interactive course materials is time consuming. Before they produce their own multimedia course materials, instructors need sufficient training on Web page design and on the use of video and audio producing techniques. According to Brown's (1998) analysis, an online class could take 40% to 50% more preparation time. This situation can be even worse when instructors prepare online technology-based courses.

Instructors are also required to keep themselves up-to-date constantly with new technologies. To make sure that students learn the latest technology, instructors need to frequently update their course content and to upgrade their online teaching labs. To keep up with the fast changing IT industry, instructors need to spend time participating in related conferences and in professional development training. It takes much more time for an instructor to help students solve their problems. Due to lack of face-to-face contact, sometimes, it may take a dozen e-mail exchanges for an instructor to understand what the problem is.

More Effort Needed from Students

E-learning is great as long as a student is well organized and highly motivated. Students need more effort to keep themselves on track. Lacking face-to-face conversations with and feedback from instructors, it is easier for students to misinterpret new concepts and theories. It also takes longer for students to learn how to use e-learning software and hardware devices. When having technical problems, students may not be able to get immediate help from technicians. After a student posts a problem online for discussion, he/she may not get immediate response from other students. All these can slow down the learning process. Kroder, Suess, and Sachs

(1998) pointed out that e-learning could take 20%-40% more time for students to learn the same amount of content as face-to-face classroom learning.

E-Learning Suggestions

From the above analysis, you can see that e-learning has its pros and cons. However, no other learning platforms can have all the advantages of the e-learning platform. The weaknesses of e-learning can be neutralized if not completely overcome. The following are some suggestions for e-learning.

- Use e-learning when it is useful. E-learning is a teaching/learning system only for appropriate circumstances.
- Careful planning is critical for a successful e-learning system.
- Instructors and students who are involved in an e-learning system need more preparations.
- E-learning requires a strong technical support team.
- An e-learning system should be supported by an appropriate budget.

Students and faculty members are the two key components of an e-learning system. We will investigate these two components. First, we will look into what types of students are ready for e-learning and what types of students are not. Then, we will consider what types of faculty members are ready for e-learning and what these faculty members need to do to meet the demands of e-learning.

Requirements for Students

E-learning is great, but it is not for everyone. As long as a student is well prepared, well organized, and has self-discipline, he/she can take full advantage of the e-learning platform. However, not all of our students have these characteristics. Most of our freshmen and sophomore students are new to the e-learning environment and they are not familiar with the course content. Certainly, these students are not the ideal population for e-learning.

Some students may not have enough self-discipline to carry out a lot of independent work. With the e-learning platform, there is no one to keep them on track. For this type of students, the chance to succeed with e-learning is not great without additional help. Some students may not have adequate e-learning skills, such as the computer hands-on skill, or typing skill. In such a case, additional assistance is needed to help these students overcome the hurdle. With the help of instructors and classmates, these students may do better in a traditional classroom.

Requirements for Instructors

Faculty members' involvement is critical to the success of e-learning. Not all faculty members are motivated to adopt the e-learning platform. Some reasons for a faculty member's resistance to e-learning are:

- Not familiar with the e-learning technology or lack of computer skills.
- Concerns about students' performance in an online course.
- Lack of time to develop online courses.
- Not enough support from administrators and computer service personnel.
- Less favorable course evaluation which may hurt their annual performance evaluation.

Also, if most of the students live in campus dorms, there is not much demand for distance learning. In such a case, administrators may not be motivated to invest additional resources in e-learning.

Faculty members tend to be motivated to make a great effort to develop online courses if there is a high demand for e-learning and strong support from administrators. For example, a strong technical support team is organized to meet the technical needs of faculty members, faculty members get teaching load reduction for developing online courses, lower course evaluation will not be used against faculty members' overall performance, and faculty members get adequate training. In such an environment, faculty members will be motivated to develop high quality online courses.

Technology-Based Courses

As pointed out in the above discussion, one of the weaknesses of e-learning is hands-on practice, especially for the students who are new to technology-based courses. This may be one of the areas that need to be improved and can be improved. Online computer labs are one of the solutions for providing more hands-on practice. A technology-based course often requires hands-on practice. Students need to practice and try things before they can become comfortable with a new technology. Online computer labs give students the practice they need and allow them to practice at anytime and at their own pace.

Before we discuss the development of online computer labs, let us first take a closer look at the technology-based courses that need computer labs to support hands-on practice.

Types of Technology-Based Courses

In higher education curricula, there are a wide range of technology-based courses that need the computer lab support for hands-on practice. Based on their functionalities, technology-based courses can be categorized as the following.

- Courses that need technology for developing multimedia course materials: Nowadays, almost all of the online courses need various technologies to develop audio, video, or life lecture course materials. A sophisticated general purpose LMS will be able to handle the technology needs of this type of course. When an online course only requires multimedia course content, usually there is no need for a special computer lab.

- Courses supported by a simulated lab for hands on practice: On a college or university campus, there are various labs for hands-on practice. Some of these labs can be simulated by computer simulation software, such as the simulated chemistry labs (Model Science Software, Inc., 2006), simulated biology labs (Virtual Courseware Project, 2006), and simulated computer networks (Lammle & Tedder, 2003). Simulated labs can be loaded on a CD/DVD or posted on the Internet. In both cases, a computer lab may not be necessary.

- Courses that are required to have hands-on practice on computer controlled mechanical or electronic instruments: For these courses, computer servers are needed to convert requests sent from Web browsers to the control language understood by the instruments. In this case, a computer lab can be just a single computer with server software installed. Such a computer lab is not a real lab in the traditional sense.

- Courses that require hands-on practice on computer software, such as the computer programming related courses: For this type of course, multiple programming software packages can be installed on a server computer. Through the Internet, students can log on to the server and work on their programming projects. For this type of course, the computer lab could be a single computer that has the programming language software installed and the remote access service implemented.

- Courses that require hands-on practice on a client-server structure, such as database systems or application development courses in the information systems curriculum: For these types of technology-based courses, if the hands-on practice is performed on a single server, we can configure students' computers as the clients and set one computer on campus as the server. If the hands-on practice is performed on multiple servers or the students are divided into multiple groups and each group has its own server, it is necessary to have a computer lab that is constructed with the client-server architecture on multiple computers.

- Courses that require hands-on practice on operating systems, or networks: The hands-on practice for these courses is often performed on multiple computers and on different networks. It may require students to reconfigure operating systems and networks. There are many such courses in the information systems, computer science, or information technology curriculum, such as courses like System Administration and Network Management. For this type of course, a computer lab is necessary. To perform hands-on practice, students have to be the administrators of the computers in the lab.

For courses in the last two categories, it is necessary to build computer labs. Before we can design computer labs for those courses, we must have a good understanding about the courses and the hands-on requirements. The following are brief descriptions of the information systems curriculum and the requirements for the computer labs by some typical courses in the information systems curriculum.

Information Systems Curriculum

The computer information systems curriculum can be categorized in two parts, the curriculum for the undergraduate degree and the curriculum for the graduate degree. Based on the curriculum recommendation for the information systems undergraduate degree programs by ACM, AIS, and AITP (Gorgone, Davis, Valacich, Topi, Feinstein, & Longenecker, 2002), the curriculum can be divided into the following areas.

- **Information systems fundamentals:** Courses in this area provide an introduction to the fields of information systems and some content on the efficient use of information technology.
- **Information systems theories:** Courses in this area cover the concepts and theories of information systems. These courses include topics such as the concept and theory about the organization of information systems, systems management, planning and strategy, quality, and decision making.
- **Information systems technology:** Courses in this area cover operating system software and networking. These courses require hands-on skills in installation, configuration, and operation of various technologies.
- **Information systems development:** Courses in this area cover information systems development. Topics such as investigating business requirements, logical and physical design, and implementation of information systems.
- **Information systems management:** In this area, topics such as information systems management, systems integration, and project management will be covered by different courses.

It is relatively easier to implement the e-learning platform for the information systems courses that do not require much hands-on practice. In this case, there are fewer requirements for networking devices, powerful servers, and high capacity network connections. The courses that do not require much hands-on practice are not very different from those in history, English, and education. Colleges and universities may have already created online courses in those fields. If so, adding this type of information systems course should be effortless. For this type of course, all they need is a well established learning management system for posting online course materials and managing students' assignments. Nowadays, most of the learning management systems will also support multimedia functionalities. With those multimedia utilities, instructors can provide video, audio, and interactive course materials to enrich online courses. Some of the courses in the information systems theory and information systems development areas are ideal for this type of e-learning platform.

The second type of courses in the information systems curriculum includes those that involve computing on the client side. For example, programming language courses, personal productivity improvement courses, Web computing courses, and database application development courses are best fitted in this category. In addition to a well-established learning management system to support the course management, this type of courses also need to provide programming software and application development software for hands-on practice. You may find this type of courses in the information systems development and information systems foundation areas. It is not very difficult to support this type of hands-on activities. We will discuss several methods that can get the job done later in other chapters.

The third type of online courses in the information systems curriculum includes those in the information systems technology, development, and management areas. Courses such as system administration, networking, and database management belong to this type. These courses present the most difficulties to the e-learning platform. In addition to a well-established learning management system and application software on the client side, it is also required that students be able to perform operations on the servers through the Internet. Because the networks of colleges and universities are heavily protected, students are prohibited to perform any administrative tasks. We still can do a few things to work around with this problem. We will discuss some solutions later in other chapters.

Courses Depending on Computer Labs

Application Design

Under this title, there could be several courses such as Application Development with GUI or Web Application Development. Most likely, these courses are taught at the junior level. Some of the higher education institutions may cover it in the senior level. These courses often cover the application development on the client side of a client-server structure. They teach students the theories of application design, how to develop forms and reports, how to configure remote access to a database, and how to consume XML data posted by Web services.

The computer lab is usually constructed on a three-tier client-server structure. It needs a server to host the database. The middle tire can be installed on the same server or use another server depending on how much workload the server has to take. Students should be able to access the server through the Internet. Students' home computers are used as client computers. The client software will be installed on their home computers. After the lab is set up, the computing environment is relatively stable. There is no frequent reconfiguration of operating systems and change of network structure, so it is relatively easy to manage this type of online computer labs.

Client application development software may include Microsoft Office, Microsoft Visual Studio, Microsoft InfoPath, SQL Server Client, Oracle Development Tools, and Linux Enterprise WS. The software for the middle tier of a client-server structure may include Apache Web Server, Microsoft Internet Information Server, or Oracle Application Server. For the database server, there is database management software such as Microsoft SQL Server, Oracle Database, or IBM DB2. Application development software can also be installed on the server computer so that students can run the software directly on the server.

Database Processing Systems

Several courses with different names may be related to this title. The kind of course is defined by the Association for Computing Machinery (2005) as a senior level database system development course in the computer science or information systems curriculum. This is a course that teaches students about database design and database implementation. It often requires students to develop a database on a client-server structure. On the server side, students will create database servers and perform database administration tasks such as database backup and restoration, and performance tuning. On the client side, students need to develop database applications such as forms and reports. Due to the fast development of the Internet technology, the subject of Web database is also covered in this type of course. Some up-to-date topics such as multitier architecture, XML, Common Language Runtime (CLR), Real Application Cluster, and business intelligence may be introduced in this course as well.

Client computers may require an operating system such as Windows XP Professional or Red Hat Linux WS to be installed. Other software may include SQL Server Client or Oracle Application Development Client and some database design and application development software such as Microsoft Visio, Microsoft Visual

Studio .NET, and Microsoft Office. On the server side, a server operating system such as Windows Server 2003 or Red Hat Enterprise Linux 4, and a database server such as SQL Server 2005 or Oracle10g should be installed. The network should be configured to support a client-server computing architecture. To enhance the group learning experience, the network in the lab should be configured to have multiple local area networks that are attached to the backbone network in the lab. Students in each group should be able to access their own local area network.

To gain the critical firsthand experience in developing and managing databases, students need to have hands-on practice on both the server side and the client side. The hands-on practice will help students understand how a database works and gain skills on database development and troubleshooting.

Telecommunication and Networks

There are several network or telecommunication courses related to this title. Depending on the content, they could be junior level or senior level courses. This type of course teaches students about the theories and application of telecommunication and networks. It may cover the concepts of transmission media, long distance transmission, packet transmission, local area network, wide area network, networking protocols, and so on. For hands-on practice, students may need to design networks, configure network operating systems and network equipment, manage networks, and implement network applications, prevent disasters and enforce security measures.

To perform the hands-on practice, a computer lab is needed. The operating system used to manage the networks can be Windows Server, Linux, and UNIX from various vendors. With a network operating system, students can configure local area networks, set up Internet connections, perform network security tasks, create routers, develop distributed computing environment, and manage the network involving multiple subnets. In the lab, students can also configure network equipment, and troubleshoot the common network problems.

System Administration

This course covers operating system management at the enterprise level. It can be taught at the sophomore, junior, or senior level. The topics may include installation of operating systems, operating system configuration, security management, file system management, system performance tuning, managing network, and configuring servers such as Web server, file server, printing server, mail server, and other servers. Students will develop a client-server structure to support the computation needs on the client side. From this course, students will also learn how to enforce corporate policies and how to manage distributed computing systems.

Various operating system software packages can be used to support this course. For example, you can use Windows Server 2003, Red Hat Enterprise Linux 4, or UNIX from various vendors. To be able to perform hands-on practice on operating systems, students must have the administrator's privilege. Similar to the network related courses, this course's demand for the computer lab is high and we should not let other courses share the same lab with this course due to the instability of the operating system. The group activities also require the network in the lab be divided into multiple subnets.

Information Security

There are several courses related to this title. This kind of course covers the concepts and practice of planning and managing the security of an information system. The topics may include:

- **Data assurance:** It may cover the concepts and practice of data integrity, data availability, and data confidentiality. It may also cover cryptography, digital signature, and other measures for providing secure data communication.

- **Network security:** It may cover issues related to discovering the sources of network attacks, identifying and responding to an intrusion, configuration of firewalls, and selection of the best security approach for various network protocols and selection of network security software and hardware.

- **Operating system security:** It may cover subjects such as identifying and applying the best security strategies for Windows, Linux, and various UNIX operating systems, applying measures appropriate to various operating systems, configuring different types of servers for safer access, and setting up user authentication.

- **Application security:** It may include security management for e-mail and Web services, implementing secure e-commerce applications, and enforcing security measures for some special purpose servers.

- **Internet security:** The issues to be considered may include how to set up appropriate security measures for Web servers, how to secure Web browsers, how to protect portals from hackers, and how to enforce security measures to protect computers from virus threats such as spam, worms, and spyware.

- **Security policies:** Subjects under this topic may include setting up security policies for an organization, devising a data protection plan, assessing the costs and benefits of various security policies, educating the organization's leaders on the seriousness of security problems, and evaluating the organization's security measures.

As seen above, there are various topics in an information security course that need the support of a computer lab. Students in the class may be divided into groups and each group should have its own network. Students should have the administrator's privilege to allow them to enforce security measures, test their server and network, and interpret the results. The security enforced on the computers in the lab may block other students' access to the server. Therefore, the network used by the information security course should be isolated from other courses.

Final Thoughts on Computer Lab-Dependent Courses

It is beyond the scope of this book to describe every course that requires computer labs in the higher education curriculum. Those of them mentioned above will be used in this book to illustrate strategies and technologies in developing online computer labs. These courses will be used as case studies in later chapters when we design computer labs and develop course materials. We will use these courses to demonstrate the development of the server and client for a computer lab and lab resource sharing for multiple online courses.

Challenges for Teaching Online Technology-Based Courses

The progress in developing learning tools and course management tools has motivated universities and colleges to accept the Web-based teaching as one of the teaching and learning platforms. Web-based training is popular in IT industry. Employees use WBT to prepare for various certification exams. WBT is also used by companies for employee development training, new technology training to keep up with the current trend, and service training to ensure customer satisfaction. According to The Center for Online Learning (2000), in the year of 1999, "92% of large organizations were implementing some form of e-learning". WBT has also been implemented by colleges and universities worldwide.

Although many courses are online now, only a few of them are supported by real-life online computer labs for hands-on practice. Many colleges and universities are still having difficulty supporting fully online technology-based courses, especially the courses that need computer labs for hands-on practice. One of the reasons is that a general purpose LMS is not able to support hands-on practice on IT products and electronic instruments. The second reason is that the technical support team may not know how to meet the special needs of technology-based online courses. The third reason is that WBT of technology-based online courses are still new to faculty members and students, who are not familiar with the technologies used to develop online computer labs. These faculty members and students need to improve themselves before they can say yes to the lab-based online courses. Another challenge is lack of funding to support WBT, especially for small universities that do not have enough funding and technical support. Among many challenges, we will discuss some of them that are closely related to the development of online computer labs for technology-based courses in the following.

Challenges in Technology

Technology is the driving force of WBT and it also presents some potential challenges to WBT. Technical difficulty is one of the challenges that prevent higher education institutions from offering online technology-based courses. The survey conducted by Kruse (1999) indicated that 40% of the learners had problems with technology.

In the LMS supported courses, the technologies may include audio, video, virtual classrooms, chat rooms, discussion groups, Web authoring tools, and Web programming languages. Although, most of these technologies are well supported by some LMSs, it may still be a challenge for faculty members and students to learn how to use these technologies. The technologies involved in the online computer labs are much more complicated than those of the LMS supported courses. For an online computer lab, in addition to the above technological challenges, faculty members and students also need to handle the network technologies, client-server architectures, server technologies, remote accessing technologies, distributed computing environment, Web service related technologies, and Internet security issues. The challenges may vary from one campus to another. In the following, let us examine some possible technological challenges that an online computer lab may present:

- **Computer systems used as servers:** To handle multiple simultaneous online sessions for hands-on practice, servers are required to have adequate speed and memory. To reduce performance bottleneck caused by I/O and to improve reliability, servers may need to be configured as a grid system. For better performance and reliability, multiple hard drives need to be combined as a RAID system. A computer lab may need to support multiple operating systems. There may be a requirement for a network-based storage device such as Storage Area Network (SAN) to share the data (Chevance, 2005).

- **Personal computers for clients:** One of the technical challenges is that computers on the client side are different. The client computers may run on different operating systems and have different configurations on memories, hard drives, and peripheral equipment such as network cards, CD/DVD drives, and monitors. It is difficult to require all of the client computers to run the same type of client software. Also, some of the client computers may not be powerful enough to handle online hands-on practice and some of them may not have the proper audio and video equipment to run multimedia files.

- **Internet connection:** Hands-on practice over the Internet often requires a high-speed Internet connection. Some students may not be able to pay for the cost of the high-speed Internet connection. Also, the high-speed Internet connection may not be available in some rural areas (Zhang & Wolff, 2004). Another challenge is the stability of the Internet connection. In some areas, the DSL service may significantly slow down or may even be unavailable from time to time. Some unpredictable events may stop the Internet service for a significant long period of time. These unfortunate events can cause situations in which students are not able to complete their assignments on time.

- **Software:** There are many types of software involved in supporting technology-based online courses. Students and faculty members need to know how to use the software such as LMS software for course management, Web authoring software for developing multimedia teaching materials, collaboration software for interactivities, Internet related software such as the Web browser and Web server, and software that supports Web conferencing, virtual classrooms, intelligent tutoring systems, and virtual computer labs. In addition to the above software required for general technology-based online courses, it is also required that the faculty and the support team know how to use the server software supporting computer labs and Web sites, application software required by technology-based courses, and know to how to configure remote access so that students can access the server. The installation and management of the software on each tier of the client-server architecture is not only time consuming, but also requires a wide range of knowledge on configuring and using these software (Goodyear, 1999).

- **Hardware:** Installing and configuring hardware devices is also a challenge for faculty members and students. Faculty members who want to develop multimedia teaching materials may need to know how to use various multimedia hardware equipment such as digital cameras and camcorders for capturing images, Web conferencing cameras, speakers, and microphones. For faculty members and support team members who are in the online computer lab development team, they need to know how to install and configure hard drives, video cards, memory chips, printers and scanners, monitors, optical storage devices, network devices, and uninterruptible power supply (UPS) units. From time to time, they need to look for drivers to match new hardware installed on the system. They may also need to troubleshoot problems of hardware devices and replace malfunctioning devices and parts.

- **Security:** The support team and faculty members who are involved in online lab development need to know how to install and configure firewalls, install and update antivirus software, and remove virus from infected computers. Students need to properly configure security properties for their Web browsers so that they can communicate with the remote course-supporting server through their Web browsers.

Before an online computer lab can be used to support technology-based online courses, the support team and faculty members must confront and overcome the technological challenges. They need to try their best to make everything work properly before the first day of class.

The development of technology is so fast that the support team, faculty members, and students will encounter new challenges every semester. This requires all the people who are involved in WBT to continue to update their personal and collective skills and knowledge to keep up with the development of technology. It takes one's passion and commitment to face the challenges of technology.

Challenges in Resources

The design and development processes of WBT need to be supported administratively and financially. As pointed out by Ruth (2006), institutions must provide appropriate resources to address the needs of WBT.

- **Administrative support:** In her report, Pirani (2004) emphasizes that administrative leadership should play a strong role in e-leaning. It is important to build a common vision about the role of e-learning in an institution and promote it from top down. A common vision provides a communication platform where administrators and instructors can build a consensus and create a set of e-learning expectations. In deed, it is difficult to get financial support and technical support without the support from administrators. A WBT development process needs the supports from various resources. Among them, the administrative support is the most important one.

- **Financial support:** To develop an online computer lab, we need software, hardware, the Internet service, and a technical support team. It is doable if the online computer lab project is supported by adequate funding to purchase the software and hardware, and to pay the support team. However, funding is not always available, especially, for smaller colleges and universities. Many small colleges and universities are short on financial support for the fast changing computer technology related courses. The price for a fully functioning medium-sized online computer lab can be over a quarter million dollars. Obviously, some of the small colleges and universities are not able to handle that.

- **Room support:** Even though it is an online computer lab, it still needs a secured room to keep the servers, network equipment, and other hardware. The room should have reliable power supply and be wired so that it has the access to the main network on campus and the Internet.

- **Multiple labs:** As mentioned earlier in the descriptions of courses, some of the courses need their own labs and others can share the same lab with other courses. We may have to develop multiple labs and each serves a different purpose. This means that the cost of online labs could be several times higher.

- **Workload:** Due to the shortage of men power, many colleges and universities may not be able to form a technical support team to support online computer labs. Often, the lab design and maintenance become faculty members' job. It is hard for faculty members to take a full responsibility of the online computer labs since they have to develop online teaching materials, grade assignments, do research, keep up with the current trend in their teaching fields, and serve on committees of their universities. In addition, faculty members must be involved in the development of online computer labs so that the labs can meet the requirements of the ever changing teaching plans. The work load is a big concern for faculty members. To have successful WBT in the long run, higher education administrators need to find a way to reduce faculty members' workload.

Challenges in Technical Support

In their writing, Arabasz and Baker (2003) point out that the technical support issues for faculty and students must be addressed since all the means of developing efficient e-learning are related to meeting the technological requirements of faculty and students. They find that, when facing new technologies, faculty members need the most technical support due to lack of knowledge in designing courses with technology. On the other hand, students need to be encouraged to use e-learning technology more by providing them with "an adequate network infrastructure" (Arabasz & Baker, 2003, p. 4). Higher education institutions need to keep up with technology to meet the technological needs of students and faculty members.

Arabasz and Baker (2003) also point out that "instructional technology support for pedagogical issues is a priority at many institutions" (p. 4). Instructional technology support is especially important when implementing online computer labs. As pointed in the course descriptions, faculty members and students have to be administrators for the computers in the labs. Technical support from the computer service team is not just to help the front-end users to log on to the computers or use e-mail, it is required to support the designing, implementing, and maintaining of the online computer labs. The support needed by faculty members and students is given below:

- Working with faculty members to design computer labs to meet requirements of hands-on practice in various technology-based courses. The hands-on requirements may vary from one course to another. The computer service team needs to have a wide range of knowledge about various technologies.

- The implementation of online computer labs should be ahead of class schedule to leave enough time for faculty members to develop online teaching materials.

- Configuring lab computers to allow students to log on as administrators through the Internet at the same time prevent students from accessing other computers on campus.

- Providing training for faculty members and students in using the equipment in the online computer labs and assisting faculty members in creating lab manuals and online teaching materials.

- When instructors are developing online teaching materials and students are working on homework assignments, they may run into technical difficulties. In such cases, technical support is much needed.

- Managing online computer labs. Ideally, the online computer labs should be available to faculty members and students 24 hours a day and 7 days a week. To keep the online computer labs running like that, we need a great deal of support from the computer service team. The computer service support team will perform maintenance tasks such as system backup and restoration. After each semester, all the computers in the online computer labs should be re-imaged to get ready for the next semester.

- Restoring the server or even the entire online computer lab after an unpredictable event such as a power outage, hurricane, earthquake, and problems due to improper operations. Being a system administrator, a student or faculty member may misconfigure the system and cause the system to crash. If that happens, system restoration needs to be done by the support team.

- Detecting and removing viruses. A lot of remote accessing software requires some of security measures to be removed so that client-side computers can run ActiveX or other plug-ins. This may open the door for Internet viruses such as spam, worms, and spyware. Technical support in this area is certainly crucial for the success of teaching technology-based courses online.

- Recommending and purchasing hardware, software, and other technology equipment used by online computer labs.

Although we all know the importance of technical support, in reality the much needed support may not be available, especially at small colleges and universities. It is really a challenge to develop online computer labs without a devoted technical support team. Often, on a small college campus, the whole computer service team has only one or two technicians. As you can imagine, the support for online computer labs from such a team will be very limited. In such a case, the technical support tasks are shared by faculty members and the computer service team.

In this book, we will deal with some of these challenges to make sure that technology can keep up with the requirements of technology-based courses. In later chapters, some recommendation of solutions will be given to help readers to overcome some of the challenges in developing real-life online computer labs.

Online Computer Lab Development Framework

An online computer lab development framework consists of several stages including planning, designing, implementing, managing, and testing. To help readers to follow the discussions on the issues throughout the book, a framework that consists of the stages of online computer lab development is described in the following. The description of each stage includes the issues and problems associated with that part of the lab development. Information on the coverage of these stages in each chapter is also given.

The development of an online computer lab is heavily dependent on the technologies. These technologies allow students to remotely access online computer labs and to perform server-side hands-on practice. It is beneficial for technicians, faculty members, administrators, and whoever is involved in the online computer lab development process to understand what the technologies can do and how we can use them to support hands-on practice for technology-based courses. Chapter 2 will briefly describe the roles of the technologies commonly used in developing online computer labs. After discussing the roles of the technologies, we will exam which type of WBT requires online computer labs to support hands-on practice.

The first stage of the online computer lab development process is to decide what type of online computer lab should be developed to meet the requirements of technology-based courses. To achieve this, the lab development team needs to investigate the needs of hands-on practice, technical support, cost, and the complexity of the technologies. Chapter III provides readers with a systematic way of identifying the needs of an online computer lab. It will discuss the topics such as identifying hands-on requirements, identifying resources, and assessing costs. This chapter will also discuss the issues such as budgeting, scheduling a timeline, forming a project development team, and implementing the project.

After the hands-on requirements of a technology-based course are identified, the next stage of the development is to design a lab that will meet the design objectives. Chapter IV will cover the issues related to the logical design and physical design. The first task in a design process is to create a logical model representing the online computer lab. Then, the model is used to verify that the objectives have been achieved. In this chapter, various online computer lab models are investigated. This chapter will also discuss the issues related to the physical design, including the strategies to select technologies to be used for the lab construction. The hardware and software requirements will be specified for various technology-based courses.

After the online computer lab model is selected and the technologies are specified for the components in the online computer lab, the next development stage is to implement the computer lab by putting the technologies together to construct the lab physically. Often, an online computer lab is constructed on client-server architecture. Chapter V will address the issues of server implementation. Servers play a crucial role in supporting the daily operations of an online computer lab. Chapter V will discuss the configuration issues for each building block of the server architecture. The issues related to the communication among servers and the issues related to server deployment are also discussed in this chapter.

After servers are developed, our next task is to develop networks that will connect the servers to the client computers. Chapter VI deals with the issues related to the implementation of networks for each type of online computer lab model. In Chapter VI, we will investigate how network equipment is used in an online computer lab and how to configure the networks to meet the teaching and hands-on practice requirements.

Once the implementation of servers and networks has been completed, it is time to configure client computers so that they can be used to access the servers. In Chapter VII, we will discuss the issues related to the configuration of the client side. We will also discuss the topics related to the configuration of client hardware and software. This chapter will deal with the implementation of remote access. The computers on the client side need to handle multimedia materials posted on the servers. Therefore, the discussion about the installation and configuration of collaboration tools, and of audio and video devices is included in this chapter as well.

At the stage of online computer lab implementation and the stage of lab management, a very important task is to protect the online computer lab from hackers and hundreds of malicious viruses. In Chapter VIII, we will deal with some security issues and come up with some solutions to security problems. We will identify potential vulnerabilities that may exist in an online computer lab. We will also deal with issues related to security policies, security measures, and security tools used to implement security measures.

Before an online computer lab is ready for online teaching and hands-on practice, lab manuals and other online teaching materials must be properly developed and made available to students. The development of online course materials is also a task that belongs to the lab implementation stage. In Chapter IX, we will discuss the issues related to the design of course materials, the use of multimedia course content, and the technologies used in developing course materials. This chapter includes a case study to demonstrate the process of developing lab-based teaching materials.

After an online computer lab is implemented, the next stage in the online computer lab development process is the lab management stage. To support the hands-on practice required by technology-based courses, it is crucial to keep an online computer lab fully operational. Chapter X deals with issues related to online computer lab management. For an online computer lab that is designed to support technology-based courses, the lab management is a challenging task due to the fact that students and faculty members must be given the administrator's privilege. In this chapter, we will first discuss the issues related to daily maintenance of the online computer lab. Then, we will investigate some of the lab management tools and come up with some general procedures for troubleshooting. This chapter will also deal with the issues related to system backup plans, system backup tools, system recovery, and system performance tuning.

An online computer lab development process is not a linear process. It is a cycle of designing, implementation, testing, and evaluation. It is difficult to make everything perfect at the first time. Therefore, testing and evaluation are important stages for improving an online computer lab. In Chapter XI, we will discuss the issues related to testing and evaluation of an existing computer lab. We will first discuss the testing of hardware, software, network equipment, remote accessibility, and course content. Then, we will discuss lab evaluation issues. We will investigate various procedures of carrying out testing and evaluation. One of the evaluation tasks is to analyze the effectiveness. Issues related to effectiveness will also be discussed in this chapter.

Web-based teaching theories and technologies are changing rapidly. In Chapter XII, we will investigate the theories and technologies that have the potential to make an impact on Web-based teaching and online computer labs. We will first discuss the trends in Web-based teaching, such as future e-learning structure, e-learning management, and e-learning course content development. Then, the trends in lab development technologies will be explored. We will examine the trends of the software, hardware, and network technologies, which are potentially useful for the development of an online computer lab.

Conclusion

This chapter has provided an overview about Web-based training and Web-based teaching. It has reviewed the short history of e-learning. It described the role of Web in e-learning. After introducing the Web-based teaching systems, the chapter analyzed the advantages and disadvantages of Web-based teaching. This chapter explained the advantages of Web-based teaching in terms of its availability, flexibility, cost efficiency, maintainability, and learning enhancement. It also dealt with the disadvantages of Web-based teaching. We can overcome many of the disadvantages by building properly designed online computer labs. To help readers to better understand the design objectives in later chapters, this chapter gave some brief descriptions of some commonly offered technology-based courses in an information systems curriculum. To prepare readers to get ready for their online computer lab projects, this chapter listed challenges an online computer lab developer may encounter. The challenges were discussed by looking at the categories such as technology challenges, resource challenges, and technical support challenges. By knowing these challenges, an online computer lab developer can plan ahead and avoid the pitfalls that so many developers have encountered in implementing Web-based teaching.

The biggest advantage of Web-based teaching is that it can reach out to students who are unable to attend classes in a traditional classroom setting or who want an alternative to the classroom. Online learning allows students to control how and when they learn. No other teaching platform has this advantage. It is worth it to make extra effort to develop Web-based teaching due to its flexibility and the fact that it can increase enrollment and save money in the long run for higher education institutions.

References

Arabasz, P., & Baker, M. B. (2003). Evolving campus support models for e-learning courses. ECAR, pp. 1-9. Retrieved March 12, 2007, from http://www.educause.edu/ir/library/pdf/ERS0303/ekf0303.pdf

Association for Computing Machinery. (2005). Computing curricula: Information technology volume. Retrieved March 12, 2007, from http://www.acm.org/education/curricula.html

Brown, B. M. (1998). Digital classrooms: Some myths about developing new educational programs using the Internet. T.H.E. Journal. Retrieved March 12, 2007, from http://thejournal.com/articles/14064

The Center for Online Learning. (2000). Online learning is fast becoming the standard. coLearn. Retrieved March 12, 2007, from http://www.colearn.com/aboutoll.html

Chevance, R. J. (2005). *Server architectures: Multiprocessors, clusters, parallel systems, Web servers, storage solutions.* Burlington, MA: Digital Press.

Chu, K. C. (1999). The development of a Web-based teaching system for engineering education. *Engineering Science and Education Journal, 8*(3), 115-118.

Driscoll, M. (2002). *Web-based training: Designing e-learning experiences.* San Francisco: Pfeiffer.

Driscoll, M., & Carliner, S. (2005). *Advanced Web-based training strategies: Unlocking instructionally sound online learning.* San Francisco: Pfeiffer.

Goodyear, M. (Ed.). (2000). *Enterprise system architectures: Building client/server and Web-based systems.* New York: CRC.

Gorgone, J. T., Davis, G. B., Valacich, J. S., Topi, H., Feinstein, D. L., & Longenecker, H. E. J. (2002). *IS 2002: Model curriculum and guidelines for undergraduate degree programs in information systems* (Report). ACM, AIS, and AITP. Retrieved March 12, 2007, from http://www.acm.org/education/is2002.pdf

Hall, K. A. (1970). *Computer assisted instruction: Status in Pennsylvania.* Bureau of Educational Research, Pennsylvania Department of Education.

Hall, B. (1997). *Web-based training cookbook.* Hoboken, NJ: John Wiley & Sons.

Horton, W. (2000). *Designing Web-based training: How to teach anyone anything anywhere anytime.* New York: Wiley.

Kozma, R. B. (1994). Will media influence learning? Reframing the debate. *ETR&D, 42*(2), 7-19.

Kroder, S. L., Suess, J., & Sachs, D. (1998). Lessons in launching Web-based graduate courses. T.H.E. Journal. Retrieved March 12, 2007, from http://thejournal.com/articles/14089

Kruse, K. (1999). *Real world WBT: Lessons learned at the Fortune 500.* Paper presented at the ASTD International Conference & Exposition. Atlanta: The American Society for Training & Development.

Lammle, T., & Tedder, W. D. (2003). *CCNA virtual lab, platinum edition (Exam 640-801).* Alameda, CA: Sybex.

Lee, W. W., & Owens, D. L. (2004). *Multimedia-based instructional design: Computer-based training; Web-based training; distance broadcast training; performance-based solutions* (2nd ed.). San Francisco: Pfeiffer.

Model Science Software, Inc. (2006). Chemistry lab simulations for the classroom, the laboratory and the Internet. Model Science Software. Retrieved March 12, 2007, from http://modelscience.com

Nott, M. W., Riddle, M. D., & Pearce, J. M. (1995). Enhancing traditional university science teaching using the World Wide Web. In I. Selwood, P. Fox, & M. Tebbutt (Eds.), *Conference abstracts of WCCE95, the Sixth IFIP World Conference Computers in Education* (p. 12). Birmingham, UK: Aston University.

Oblinger, D. G., & Hawkins, B. L. (2005). The myth about e-learning: "We don't need to worry about e-learning anymore." *EDUCAUSE Review, 40*(4), 14-15.

Pirani, J. A. (2004). Supporting e-learning in higher education. ECAR. Retrieved March 12, 2007, from http://www.educause.edu/ir/library/pdf/ERS0303/ecm0303.pdf

Ruth, S. R. (2006). E-learning: A financial and strategic perspective. *EDUCAUSE Quarterly, 29*(1). Retrieved March 12, 2007, from http://www.educause.edu/apps/eq/eqm06/eqm0615.asp

Sloan Consortium. (2005). Growing by degrees: Online education in the United States, 2005. Retrieved March 12, 2007, from http://www.sloan-c.org/publications/survey/survey05.asp

Szabo, M., & Montgomery, T. (1992). Two decades of research on computer-managed instruction. *Journal of Research on Computing in Education, 25*(1), 113-133.

Virtual Courseware Project. (2006). Biology labs on-line. Retrieved March 12, 2007, from http://vcourseware.calstatela.edu/BLOL

Zhang, M., & Wolff, R. S. (2004). Crossing the digital divide: Cost-effective broadband wireless access for rural and remote areas. Retrieved March 12, 2007, from http://www.coe.montana.edu/ee/rwolff/Divide-rev4.pdf

Chapter II

Web-Based Teaching Systems and Technologies

Introduction

In Chapter I, we give a brief overview about Web-based teaching and technology-based courses. The discussion in Chapter I indicates that Web-based teaching is heavily dependent on technologies, especially for technology-based courses and extra effort is needed to implement Web-based teaching for technology-based courses. As technology advances, even those technology-based courses that depend on a computer lab can be taught 100% online. Students are able to access the computers specially designed for their courses via the Internet. Technology tools have been assisting us in achieving this goal. It will be beneficial for the development of an online computer lab if we can better understand the functionalities of the technologies involved and the role played by these technologies. In this chapter, we will discuss the issues related to the technology tools that are commonly used to create online teaching materials, manage Web-based teaching (WBT) systems, and develop online computer labs. An overview will be given about what the technologies can do and how we can use them in creating online teaching materials. After those technologies are introduced, we will exam the types of Web-based teaching systems that require online computer labs to support hands-on practice.

Background

The early Internet-based teaching was based on main frame computers, and mono-chrome and text-only monitors. A main frame computer was used as a server, and a monochrome and text-only monitor served as a server-accessing interface. Due to lack of interactivity, and lack of audio and video functionalities, the Internet was used by educational institutions only as a knowledge distribution tool. Such a system is a text-only computer-based training system (Perry, 2000).

With the advance of personal computers, a computer-based training system was able to deliver a course with a graphical user interface (GUI) for interactivity and multimedia functionalities. Course materials could be burned on a CD and installed on a personal computer. This type of computer-based training system could be used to teach various subjects, from children's books to technology-related training (Kruse & Keil, 2000). Nowadays, many of the computer professional development related books include the trial version of the software and some computer-based training materials such as programming code for the examples included in the books. Some books even include virtual labs for hands-on practice. For example, some of the books that prepare readers for taking the Cisco Certified Network Associate exam often include a simulator that simulates a virtual network (Lammle & Tedder, 2003). As another example, the virtual lab included in some of the books used for preparing Microsoft Certified Systems Engineer and Systems Administrator exam allows readers to perform some essential administration and management tasks (Sheltz & Chellis, 2002).

The creation of the World Wide Web has extended computer-based training to students anywhere at anytime. Many new technologies such as digital audio and video equipment, Web conferencing tools, high volume portable storage devices, much faster and more powerful personal computers, and various learning management software packages are now available to support Web-based teaching. The advance of these technologies pushes Web-based teaching to reach out to more students. In the field of employee training, WBT supports online conferencing, virtual reality, and live information.

For teaching technology-based courses online, a Web-based teaching system is the desired platform. It can support multiple courses simultaneously. Multiple users can log on to the same WBT system at the same time. This allows group activities even though the group members may be thousands of miles apart (Dara-Abrams, 2002). A Web-based teaching system is also good to support multiple software packages for hands-on practice. For many technology-based courses, hands-on practice may require multiple software packages. For example, a database system development course may need software for the front-end, mid-tier, and back-end. These software packages can all be installed in an online computer lab and can be accessed by students through the Internet. Therefore, an online computer lab that is part of a Web-based teaching system can take all the advantages of a Web-based teaching system.

Higher education institutions have been actively involved in the research of Web-based learning. To improve Web-based teaching, they have done many research studies. For example, in 1997, The Oncourse Project at Indiana University developed the template-based course management system, which was later used by Blackboard, WebCT, and other course management systems (Wikipedia, 2007). Similar effort was made by Veglis (2000), in whose lab some Web-based interactive tools were developed. In the field of control engineering, Valera, Díez, Vallés, and Albertos (2005) report that a MATLAB Web server was used to simulate dynamic processes and allow students to perform hands-on practice required by process control courses. One of the books that address online lab issues is the book edited by Fjeldly and Shur (2003). This is the first book that deals with running real solid state electronics experiments via the Internet for the semiconductor related courses. Later in this chapter, we will discuss the early online labs in detail.

As mentioned in Chapter I, a LMS is not adequate to meet hands-on practice requirements of technology-based courses. Especially, it does not allow the practice on server-and-network administration. Among a large number of books and articles about LMSs, however, there are very few about online teaching of technology-based courses in computer information systems or related curricula. In their article, Rowe and Gregor (1999) discuss the implementation of Web-based programming courses and Web-based tutoring. The CGI script behind the Web browser was used to compile and execute the C++ code submitted by students. A simple online lab was created to allow students to experiment with their programming code. The disadvantage of such a system is that it is not flexible enough to meet the requirements of today's technology-based courses. When a programming language is updated with new functionalities or a current programming language is replaced by another programming language, one has to rewrite the CGI scripts. Also, if other universities want to implement a similar project, they may have to write their own code. Such a lab will not allow students to directly access the server or network. As technology advances, most of these difficulties can be resolved by using remote access tools. In later chapters, more detailed description about the remote access tools will be given.

Correia and Watson (2006) report that the virtual network technology is used in a computer lab on their campus to support hands-on practice required by a networking course. Their study is significant since only a few such research studies can be found in this area. Their lab is a local computer lab on campus. In this chapter, several online computer labs will be investigated. It is the main objective of this book to give a systematic study of online computer labs that support the hands-on practice of various technology-based courses.

The hands-on practice of a technology-based course in the computer information systems curriculum often requires more sophisticated operations such as server-and-network administration or developing a network-based database on client-server architecture through the Internet. According to the author's knowledge, there has not been a book that is specifically about online computer labs for teaching technology-based courses. In this book, we will discuss the design strategies, implementation methods, and the effectiveness of online computer labs that meet the hands-on requirements by the technology-based courses. Before we start the online computer lab development process, we need to know what technologies are used in a Web-based teaching system.

Technologies for Online Teaching

The development of online computer labs involves various technologies in the developing process. The technologies are used to:

- Generate multimedia teaching materials.
- Develop servers which make the teaching materials available to students.
- Connect students' computers to the servers.
- Create and manage online course Web sites.
- Implement Web-based computer labs.
- Let students browse multimedia course content and perform hands-on practice.
- Allow students to take online exams and submit their homework and projects.
- Grade exams and other assignments.

We will briefly discuss some common technologies used to accomplish the above tasks in this topic. We will investigate the functionalities and capabilities of these technologies. First, let us take a look at how computers in an e-learning system are connected together. Figure 2-1 gives an illustration of the components in a Web-based e-learning system.

In a Web-based teaching system, computer systems can be categorized into two types based on their functionalities. If a computer system is used for user management, network management, security management, hosting and deploying applications such as database server and multimedia course materials, it is called a server. A server allows other computers in the same network to share data, files, printers, and application software. Multiple users can log on to a server simultaneously.

Figure 2-1. Illustration of Web-based learning system

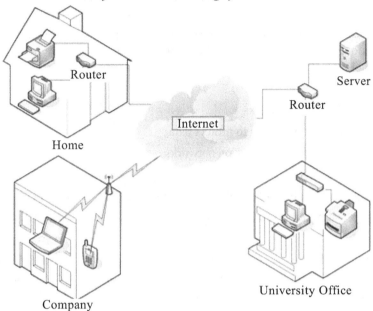

The other type of computer system can be a desktop computer, a notebook computer, or a PDA which can be used to access the server, run application software, and process data downloaded from the server. It is called a client computer. Usually, a server computer is equipped with one or more faster central processing units (CPUs), larger storage devices, and better network capabilities. On the other hand, a client desktop computer or notebook computer is often equipped with better multimedia equipment such as a faster video card, a better sound card, or a faster DVD drive.

Server Technologies

For a Web-based teaching system, technologies are needed to assist the management of classes, users, security, and networks. The technologies used to accomplish these tasks are normally installed on a server computer. The server technologies allow multiple client computers to share printers, files, or applications and to run special purpose servers such as database servers and Web servers.

Server Operating Systems

First, a server must have an operating system that is capable of accomplishing the above mentioned tasks. For a Web-based teaching system, server operating systems should be able to perform the following tasks:

- Manage and audit networks.
- Enforce security measures to protect data.
- Manage users and computers on networks.
- Store course materials and make them available to students.
- Communicate with other servers to share the distributed resources.
- Host special purpose servers or other application software.

Many technology-based courses depend on the special purpose servers which are supported by server operating systems. The following are brief descriptions of those special purpose servers:

- **Domain server:** Domain is a collection of computers, users, and network equipment using the same security model (Minasi, Anderson, Beveridge, Callahan & Justice, 2003). A domain server manages all operations in a domain. This kind of server is responsible for handling all accounts in a domain. All user names and passwords in a domain are kept on the domain server.
- **Print server:** This type of server is used to run network printers. A network printer can be shared by multiple computers on a network.
- **File server:** A file server has files or folders that can be shared by network users.
- **Application server:** An application server is used to handle business logics and access of application software. This kind of server allows multiple users to share the application software across the network. Web servers are sometimes considered as application servers.
- **Database server:** A database server is a data provider. It also provides database services to users in an e-learning system.
- **E-mail server:** An e-mail server is used to support daily e-mail operations.

The commonly used server operating systems are Windows Servers which, in 2005, held 36.9% of the market share on all server revenue. Various versions of UNIX systems had 31.7% of the market share, and Linux had 11.5% of the market share (Lawson, 2005). The following are descriptions of these servers:

- Windows-based servers are often used for network and security management, and hosting applications such as database servers, ERP (enterprise resource planning), e-mail servers, file servers, Web servers, mail servers, VPN servers, print servers, streaming-media servers, and Web directory services. It is relatively easier to learn the Windows Server operating systems due to the graphical tools supported by Windows.

- UNIX servers can run application servers, e-mail servers, and large-sized database servers. UNIX delivers high levels of security, integration, flexibility, performance, and reliability that are essential for meeting the high demands of today's e-commerce. A UNIX server is also well-suited for research work due to its capabilities. It offers the performance needed to support data-intensive tasks such as storage management, relational databases, data mining, and data-intensive scientific applications.

- Linux is an open-source operating system. Similar to UNIX by having a clone of the UNIX kernel, Linux has all the features that a server should have, including the support for hosting application servers, e-mail servers, database servers, security management, network management, file sharing, and true multitasking. Like Windows Servers, Linux is often installed on personal computers (PCs) due to its versatile GUI capabilities.

In a later chapter, we will talk about strategies of selecting server operation systems.

Special Purpose Servers

Special purpose servers are software packages that provide services for a specific application such as database or Web. The descriptions of some special purpose servers are given below.

- **Web server:** A Web server is a software package used to provide services, such as Web service or file transfer protocol (FTP) service, to client computers. When a request for a Web page is sent to a Web server, the server identifies the name of the file containing the Web page sent in with the GET command. Then, it retrieves that file and sends it back to the browser. Sending information to whoever requests it through the Internet is risky. Web servers can enforce security measures for the serving process. Some Web servers include the user authentication component which only allows users with proper permission to access the Web pages. To protect the information being transmitted over the Internet, most of the commercial Web servers also allow an encrypted connection between a server and a client, so that sensitive information like student personal information can be transmitted safely through the Internet.

- **Virtual server:** A virtual server is a software package that simulates a computer system. It is like running another computer on your PC. It enables students to run several different operating systems including Windows Server, UNIX, and Linux simultaneously on a single physical computer. A virtual server is safe to use in a lab. If a virtual server crashes, it will not harm the host computer. The drawback of a virtual server is slow performance.

- **Communication server:** A communication server is used to enable real-time, multiway communication applications (including text, audio, and/or video). Some of the communication servers also include VoIP for online conversations and tools for recording and replaying a meeting.

- **Business process automation server:** A business process automation server can be used for automating the tasks in a business process including business process design, development, management, and monitoring. The server allows companies to use the power and benefits of Web services to collaborate their efforts in developing a dynamic distributed application system. Microsoft BizTalk Server is this kind of server. Similarly, the Business Process Execution Language for Web Services, WS-Transaction, and WS-Coordination package developed by IBM can accomplish similar tasks.

- **Virtual private network (VPN) server:** A VPN server that can be hardware or software acts as a gateway into a network or a single computer. It allows universities to use public Internet lines to create a virtual private network. By using VPN, students can access a university's private network through the Internet. Some of the operating systems such as Windows Server 2003 include a VPN server.

- **Database server:** A database server is a data provider for various database applications such as forms and reports. It can be used for database development, integration, and management. It also supports data analysis for business intelligence. Database related courses need the support of database servers for hands-on practice. The commonly used database servers are Microsoft SQL Server, Oracle Database, and IBM DB2.

It is not necessary to install all of the above servers when implementing Web-based teaching. It will depend on the online teaching requirements. In later chapters, we will address some design issues about special purpose servers.

For most online courses, the servers are managed by the computer service team. Students and faculty members work on the client side and they do not need to have the server administrator's privilege. However, for many technology-based courses, students need to perform hands-on practice on the servers. They must have an administrator account on the servers. Usually, a server is a highly prohibited area and only a few people are allowed to have the server administrator's privilege. Once the server configuration is changed, it will have an impact on all the client computers. Later, we will discuss how to resolve the conflicts.

Learning Management Systems

Servers are also used to host e-learning management software which may also be called learning management system (LMS) software. As more and more online courses are developed, universities need enterprise-level learning management system software to schedule, track, and report information on online courses. LMS is designed to handle these tasks. The tasks that should be accomplished by LMS are:

- Manage students who are taking Web-based courses.
- Manage course content, assignments, and grades.
- Post course materials to client users who, in our case, are students.
- Allow client users to upload their completed assignments for grading.
- Allow client users to take online examinations.
- Provide collaboration tools that allow students and instructors to work together.

Most university online courses are supported by general purpose LMS software packages. During the process of developing e-learning, these software packages can be used to create teaching materials and manage online courses. As online education has entered the mainstream of education, software companies and e-learning consulting companies have developed hundreds of learning management systems and Web-based training management software packages. In this paragraph, we are going to overview some of these packages. A sophisticated LMS should include the following components.

- Course development tools for developing Web-based course materials.
- Instructor tools for course planning and managing.
- Teaching tools for online help and communication between instructors and students.
- Student tools for accessing course materials, students' grades, online exams, and multimedia supports.
- Management tools for enforcing security measures, managing users, online registration, remote accessing, and crash recovery.

Almost every major LMS product includes the above mentioned components. Individual products may add additional components. Itmazi and Megías (2005) provides detailed discussion on the major LMS products such as WebCT, Blackborad, and so on. For technology-based courses, hands-on practice is a very important part of the learning process. Among the popular LMS products, there are very few functionalities that support online hands-on practice on IT products. Therefore, for technology-based courses, WebCT or a similar LMS software package alone is not enough.

Network Technologies

In the above, we have briefly discussed the functionalities of a server. An online computer lab needs networks to link servers to their clients. Next, we will take a look at local area networks (LANs). A local area network (LAN) allows computers to communicate locally. In the same LAN, computers can share files and printing equipment. Network devices used in a local area network may include network interface cards and switches. Network transmission protocols such as TCP/IP to transmit data over a network are needed. If it is a wireless local area network, an accessing point is needed. In brief, let us see what we can do with some network devices:

- **Network interface card (NIC):** It is a network device plugged into a computer and physically connects the computer to a network. A NIC can be used to send data to and receive data from a network, and can be used to detect network errors.
- **Switch:** A switch is used to connect several computers. It serves as a distribution center. When a computer sends information to another computer on the same network, the switch determines which computer should receive the information.
- **TCP/IP:** It is the combination of the Internet protocol (IP) which defines the format of data packets to be transmitted over the Internet and the transmission control protocol (TCP) which is used to control data transmission, detect data transmission errors, and retransmit the data if error occurs during the data transmission process.

To transmit data over the Internet, you will need to connect a local area network to the Internet. Since the Internet and local area network are different types of networks, you use a device called a **router** to connect them. A router can be used to forward data from one network to other networks according to the rules specified in a built-in routing table.

To communicate with the server on campus through the Internet, students need to find an Internet service provider (ISP) which provides the telecommunication transmission devices and services to allow the students remote access to the server on campus through the Internet. ISP companies charge users for their services.

Depending on the type of service subscribed by students, home computers can be connected to the Internet through a dialup modem, asymmetric digital subscriber line (ADSL), cable modem, satellite system, or WiMAX. Descriptions of these technologies are given below.

- **Dialup modem:** A dialup modem is a device used to encode digital signals generated by a computer in a carrier wave which can be transmitted across a dialup phone connection. The dialup modem is connected to a telephone. Through the telephone, a student can dial a local Internet service provider to access the Internet. Due to the low bandwidth and high noise-to-signal ratio of the telephone line, the dialup modem has a transmission rate lower than 56K bits per second, which is the lowest when compared with other Internet connection technologies mentioned above. However, it is the least expensive one.

- **ADSL:** ADSL is a technology used to deliver digital signal at high-speed over an existing telephone line. ADSL has a higher download transmission rate than the upload transmission rate. Its download transmission rate ranges from 256Kbps to 3Mbps and its upload transmission rate ranges from 128Kbps to 768Kbps. In addition to high transmission throughput, another advantage of ADSL is that, while you are connected to the Internet, you can still use the phone line for voice calls. To connect to the Internet, the ADSL technology uses a DSL modem which is a transceiver that consists of a simple router or a switch. A DSL modem forwards data from computers on a LAN to the Internet or vice versa.

- **Cable modem:** A cable modem device connects a computer to the Internet through a particular cable television channel. Cable modems let you send and receive data from the Internet. In general, the transmission rate of a cable modem is higher than the transmission rate of ADSL. The cable modem's download transmission rate ranges from 1.5Mbps to 6Mbps. Its upload transmission rate is from 128Kbps to 996Kbps. When compared with ADSL, the cable modem has a benefit for Internet access because its performance does not depend on the distance between home computers and the central cable office.

- **Broadcast satellite system:** A satellite system has a better transmission rate than that of a dialup modem. The advantage of a satellite system is that it can reach rural areas where ADSL or cable cannot reach. The disadvantage is that it is more expensive and slower than ADSL and cable. It also needs a phone line for data uploading.

- **WiMAX:** As an alternative wireless solution, one can use WiMAX which is short for Worldwide Interoperability for Microwave Access. It is a wireless technology that provides high-throughput broadband connections over a long distance. Its transmission rate can be up to 155Mbps and it can cover an area as large as a metropolitan area. WiMAX works like a cell phone system. When you turn on a computer, the computer will automatically connect to the closest available WiMAX antenna. WiMAX could potentially reach suburban and rural areas where there is no broadband Internet access. Currently, WiMAX is not widely available yet. Some analysts predicted that its extensive availability will not happen until 2010 or later; others predicted that it will happen sooner (Walton, 2005).

Through these network technologies, students will be able to view course materials posted on the server and work on a lab assignment in an online computer lab from their home computers. To successfully access the online computer lab, the computers on the client side need to be equipped with remote access technologies. Also, some multimedia hardware and software should be installed on the client-side computers so that students can process multimedia course materials. The technologies on client computers will be described next.

Workstation Technologies

To view multimedia teaching materials through the Internet and access online computer labs, a computer system on the client side should have some of the following hardware and software installed. Some of the software and hardware can also be installed on server computers.

Workstation Operating Systems

The operating systems on the client side should be able to access a remote server and run some application software required by a technology-based course. For example, instead of using the Microsoft Windows XP Home Edition, a client computer should use the Microsoft Windows XP Professional Edition. The Windows XP Professional Edition allows a user to install the Web server software, Internet Information Services (IIS), which is needed to run the application software such as Microsoft Visual

Studio.NET or Virtual Machine Remote Control. Another option is to use a Linux operating system. Some vendors provide the desktop version of Linux.

Audio

A multimedia-based lecture requires audio equipment to be installed on client computers in order to handle voice, music, and other sound. Usually, a computer has built-in audio equipment such as a headphone port, speaker port, and ports for powered and unpowered microphones. Some of the computers have low cost built-in speakers. For e-learning purposes, these devices are good enough. If a speaker and microphone are not included in a student's computer system, he/she needs to purchase them for the online courses.

It can make big difference if voice, sound effect, and music are added to a text-based lecture. In e-learning, audio can be used to explain complex concepts, and express the emphases in a lecture's notes. Digital audio comes in many different formats. The following are some commonly used digital audio file types.

- **Pulse Code Modulation (PCM):** PCM is a generic digital audio data format. It is commonly used for storing and transmitting uncompressed digital audio. It can be read by most audio applications.

- **WAV:** WAV is the default format for digital audio on Windows PCs. WAV files are uncompressed and take up a lot of space.

- **AIFF:** AIFF is the default audio format for Macintosh. AIFF format can be compressed and is supported on most other platforms and by most audio applications.

- **AU:** AU is the default format for Sun systems. The AU format can be compressed and is supported on most other platforms and by most audio applications.

- **MPEG audio:** MPEG created by Motion Picture Experts Group (MPEG) is a family of open standards used for compressed audio including MP2, MP3, and AAC.

Usually, different types of audio files are not compatible with each other. Fortunately, most of the audio players are able to support multiple formats. There are many tools available for converting digital audio to different formats. To have better sound quality, you can use the data streaming technique, which stores several seconds of data in a buffer before playback.

Video

Video equipment on a client computer includes a video cord and a monitor. The video card generates video signals and sends these signals to the monitor. The monitor displays the video signals on its screen. There are two types of monitors, cathode-ray tube (CRT) monitor and liquid-crystal display (LCD) monitor.

When video files are added to online lecture notes, it greatly improves the e-learning quality. Video conferencing can be stored in a video file and embedded in online lecture notes for reviewing. Video files can also store screens captured during a lab activity, so that they can be used as guide for hands-on practice. Like audio files, video files also have various types. The following are brief descriptions of some commonly used video files in Web-based multimedia teaching materials.

- **Motion Picture Experts Group (MPEG):** MPEG is a set of standards used for processing audio and video information in a digital compressed format. The set of MPEG standards can be divided into four parts: MPEG-1, MPEG-2, MPEG-3, and MPEG-4. With MPEG-1, you can store video signals on a CD-ROM. MPEG-1 supports a video resolution of 352 by 240 at 30 frames per second (fps), which is slightly lower than that of VCR. MPEG-2 is an extension of MPEG-1 with higher recording speed and resolution. MPEG-2 supports a resolution of 720x480. It is good enough for NTSC televisions and DVD-ROM. MEGP-3 is designed for high-definition televisions. It supports a resolution of 1280x720 at 60 fps. MEGP-3 is included into MPEG-2 as part of its TV standard. MPEG-4 is designed for transmitting images over a narrower bandwidth. It can also mix 3D objects, video, text, and other media types.

- **QuickTime (QT):** The QuickTime format is developed by Apple. It is capable of handling various formats of digital video, sound, text, animation, music, and images. QuickTime is also capable of importing and editing other formats such as AIFF, DV, MP3, MPEG-1, and AVI without requiring all media data to be rewritten after editing. With QuickTime, you can easily add movies and animations to the course content. To use QuickTime in Windows, you need to download and install the QuickTime media player on Windows.

- **RealVideo (RV):** RealVideo is widely used on desktop and notebook personal computers and mobile phones. Initially, RealVideo was designed to deliver streaming video across IP networks to personal computers at low bit rates. Now, RealVideo is widely distributed and used worldwide for high-quality audio and video over the Internet at the lowest bit rates. Data services with high-quality audio and video at low bit rates is ideal for wireless mobile devices, such as cellular phones, to receive multimedia content. RealVoice allows users to watch video on their mobile phones.

- **Windows Media Video (WMV):** As part of the Microsoft Windows Media framework, WMV is designed to handle various types of video signals. WMV files can be highly compressed. WMV can be compressed to a size that matches a specific data transmission bandwidth. Its video stream is often combined with an audio stream of Windows Media Audio. Windows Media Video 9 or later provides support for high-definition video. By using Windows Media Video 9 or later and a duel layer DVD, you can record over two hours of high-definition video.

Media Players

To playback multimedia files, it requires that a media player be installed on the personal computer. With media players, students can play the audio and video included in multimedia course materials. Media players can also be used to read the content in Portable Document Format (PDF) files. Students can use a media player with a Web browser or use it independently. When a Web browser receives a file that contains multimedia course content, the Web browser checks the HTTP header. The HTTP header contains the data types of the file. If the data type is audio or video, the Web browser finds the media player to run the multimedia file. Some of the media players can play compound media formats so that a user can play audio and video clips, movies, and games that have different formats. The following are some of the commonly used media players.

- **RealPlayer:** RealPlayer is one of the first technologies that enabled streaming video on the Web. It supports the SMIL authoring language which is used for multimedia presentations. SMIL integrates streaming audio and video with text or any other media type.

- **Windows Media Player:** Windows Media Player can be used to discover, download, organize, and play your digital media and recorded TV. It can be used for ripping CDs, burning CDs, and synchronizing video images with portable devices. It can also be used to manage media files.

- **QuickTime Player:** QuickTime Player is a component included in QuickTime that can be used to play back video, 3D objects, VR Panoramas, still images, and flash files. It can even be used for editing and authoring content for the Web. QuickTime media players support an array of media formats such as MPEG-4, MP3, and JPEG. QuickTime media players are available for both Mac OS and Windows XP.

- **Flash Player:** Flash Player can be used to play back high quality streaming video. Flash Player is a lightweight media player that can be used with the 56Kb modem Internet connection.

With a media server, you can stream video and audio data. Microsoft Windows Media Player can also burn Windows Media Audio and MP3 files.

Video Cameras

To create a video-based lecture or to have a video conference online, you will need video cameras. To be able to have a video conference with the instructor through the Internet, students need to install a video conferencing camera on their computers. The following are the descriptions of three types of video cameras:

- **Video conferencing camera:** A video conferencing camera is a small camera that can be clipped on a monitor. The video conference cameras on both computers involved in a conversation capture the images of the users and transmit them to the remote computers so that the participants can see each other.

- **Camcorder:** Another type of video camera is a camcorder which can be used to capture an entire lecture. There are two major types of camcorders, analog and digital. The images captured by a digital camcorder have better quality. It is easier to edit a digital image on a computer. For e-learning, you should have a digital camcorder. Camcorders can record the captured images to MiniDV digital videotape, flash memory, or DVD. Under low light, most camcorders produce dim and unclear video. Since most of the video-based lectures will be recorded indoors, it is important to have a video camcorder that has good low light performance.

- **Pan/Tilt/Zoom (PTZ) camera:** A PTZ camera has an optical zoom lens that gives you great focus and clarity. It also supports motion recording that allows the lens to follow moving objects such as a professor lecturing in front of a whiteboard. The captured images can be transmitted live or replayed over the Web. A high-quality PTZ camera can capture and transmit about 30 image frames per second. Some of the PTZ cameras have built-in Web servers that are used to send real-time images over the Internet. Students who participate in an online course can access the live video by using a standard Web browser.

Graphics

Graphics are often used in online teaching materials. There are two different types of graphics, vector and bitmap. The vector graphics are designed for drawings, and the bitmap graphics are designed for photographs and paintings. The Web-based vector graphics have the following formats.

- **Scalable vector graphics (SVG)**: SVG is a language used for drawing two-dimensional graphics. The files created with SVG are compact. SVG also supports scripting and animation. It is ideal to use SVG for creating interactive, data-driven, personalized graphics for online course materials.

- **Vector markup language (VML)**: VML is an XML-based language used to exchange, edit, and deliver high-quality vector graphics on the Web. VML is supported by most of the Microsoft products, such as Microsoft Office or Internet Explorer.

There are several bitmap graphics-file formats that are commonly used to develop Web-based course materials. These bitmap graphics-file formats are described below.

- **Graphics interchange format (GIF):** GIF is a common graphics format found on the Internet. GIF can handle up to 256 colors. The recent GIF product can handle up to 16 million colors. GIF is the format that is used to compress and store graphics files. GIF files are smaller and easier to transfer than most other formats. GIF is best for simple solid color graphics like clip art, logos, buttons, icons, curves, and lines. Usually, GIF is not for sophisticated images such as photographs.

- **Joint photographic experts group (JPEG):** JPEG files have better image quality. Usually, JPEG are used for photographic images or artwork. JPEG can compress and store up to 24-bit colors; this means that it can handle up to 16 million colors. The drawback of JPEG files is that it downgrades the image quality slightly when file size is reduced in order to save some space.

- **Portable network graphic (PNG):** PNG is a graphic file format supported by many new Web browsers. PNG is designed to replace the GIF format. This format is capable of handling much more sophisticated images. It is a good format for compressing and storing graphic images. Unlike JPEG, PNG is a lossless format when reducing the size and it supports up to 48 bits colors per pixel for full color images.

Animations

Animations are a sequence of images that are displayed on screen to have motion effects. For example, animations can be used to demonstrate the moving of a mouse in multimedia lab instruction materials. Similar to graphics, animations also have two basic types, vector animations and bitmap animations. These animations are described below.

- **Vector animations:** A vector animation uses vectors rather than pixels to control the motion. With a vector animation, images are displayed and/or resized using values calculated by mathematic equations. Therefore, a vector animation usually generates cleaner and smoother motions. Vector animation takes up lesser memory space, but takes more time for creating the images. One of the most commonly used vector animation programs is Macromedia Flash.

- **Bitmap animations:** A bitmap animation uses pixels to control the motion. It first generates a bitmap frame with every pixel to be displayed on a screen, next loads the frame into the system memory, and then displays the frame on the screen. Therefore, for a high quality image, a bitmap animation takes a huge amount of memory. This disadvantage has some negative effect for Web-based teaching. When transmitting bitmap animation frames over the Internet, the size of bitmap files is a major concern. One of the advantages of a bitmap animation is that it can implement complex animations such as 3D animations since it does not depend on complex mathematic equations which may not be available for very complex motion. Bitmap animations can be created by software packages such as Macromedia, Microsoft Internet Explorer, and programming languages such as C++ and Visual Basic.NET.

Web Browsers

A Web browser installed on a client computer system can be considered a graphical interface to the Internet. Through a browser, students can log on to an e-learning server, access course content, download course materials, submit homework, take online examinations, and check grades. In a Web browser, when a student enters an URL address containing the server name and the Web page file name, a name server on the network converts the server name into the IP address which is used to connect the server. The communication of a Web browser and a Web server is through a network protocol called Hypertext Transfer Protocol (HTTP), which is a set of rules for file transmission on the Internet. The request for the Web page is carried by HTTP and sent to the server. Based on the request, the server finds the Web page file and returns the content to the browser by using HTTP. Once the browser receives the content, it will read the HTML tags and format the page onto the student's screen and launch the programs that plays the audio and video attached with the Web page file. Browsers can also be used to run plug-ins, JavaScript, and ActiveX files that can be used to enhance the e-learning experience.

Most of the Web browsers are free. Students can download them from the Internet and install them on their computers. If a specific Web browser is required, students should know that before the class starts.

Web Conferencing

Web conferencing is a useful tool to assist collaboration among students and instructors. With Web conferencing, students or participants of a conference can view multimedia slide presentations through PowerPoint, see each other via video conference cameras, and use a mouse to draw or write on a virtual whiteboard included in a messenger program. There are a wide variety of Web conferencing programs. The text messaging or chat program is one of the simple Web conferencing tools. Chat provides a spontaneous exchange of information over the Internet, which hosts text-based group discussions. More sophisticated Web conferencing tools allow students exchange visual information using Webcams and streaming video online. By combining Web pages, chat, flash animation, and Internet communication technologies, Web conferencing creates an interactive meeting environment. Some Web conferencing products allow attendees to view presentations in their regular Web browsers without installing any additional software. Other Web conferencing tools also provide audio chat so that attendees can ask and answer questions through the Internet.

Collaboration Tools

Web conferencing is a great tool for collaboration. There are some other collaboration tools that are worth mentioning.

- **Intelligent tutoring system:** An intelligent tutoring system is computer software that mimics what a human teacher might do. It allows a teacher to set up a course based on their teaching experience, to monitor a student's learning activities, to raise questions, to analyze the student's mistakes, and to give hints to help the student. It can respond to students' questions and adapt to students' learning styles. With an intelligent tutoring system, instructors can track each individual student and students can learn course content in a virtual training environment.
- **Blog:** Blog is a Web page based personal diary that can be viewed by others. It is easy to set up and manage a blog. Once a blog is created, the owner can invite others to contribute their comments. In a Web-based teaching environment, blogs can be used in a number of ways:
  - Let students exchange their ideas, suggestions, and arguments on problem solving.
  - Let instructors share lecture notes with their students and keep the lecture notes from students who are not enrolled in the classes.

    ○   Invite each student to post his/her proposals for projects and let the instructor and other students give their comments on the proposals.

    ○   Promote group activities on a project assigned by the instructor.

Web Authoring Technology

Web authoring tools are software packages that can be used to create Web pages with multimedia content and link individual Web pages together to form a Web site. Some of the Web authoring tools allow users to create Web pages that include animations and interactive components. A Web authoring tool often includes a Web page editor that has the following functionalities.

•   With a Web page editor, users can create Web pages without getting into the details about HTML.

•   Most of the Web page editors support "what you see is what you get" (WYSIWYG) editing.

•   Some of the Web page editors provide a built-in spell check.

•   Web page editors allow users to create links and clickable images.

•   Users can create online forms and reports with a Web page editor.

•   Users of Web page editors can convert images from one type to another type.

Web authoring tools allow the users to specify where to store Web pages on a server and how to connect to a database which stores data used by a Web site. To improve productivity, some Web authoring tools provide templates for some special purpose Web pages such as Web forms for online registration.

Optical Drives

Optical drives are used to burn multimedia files on a CD or DVD, install application software, and play audio and video files stored on a DVD. As the size of application software gets larger and larger, a lot of the application software requires DVDs to store the software. It is necessary to have at least a DVD-ROM on a client computer so that a student can install application software. DVDs have three different types: DVD-R, DVD-RW, and DVD-RAM. Once the files are recorded on a DVD-R, the recorded files cannot be erased or modified. To delete or modify the files, your DVD drive should be capable of handling DVD-RW and DVD-RAM disks. Not all of the DVD drives on client computers have this kind of capability. In such a case, students may consider installing an external DVD drive that can handle DVD-RW or DVD-RAM disks. A dual-layer DVD drive can store up to 8GB data on a DVD. Such a DVD can hold a few hours of multimedia-based lectures. DVD-R, DVD-RW, and DVD-RAM disks can be used to carry multimedia data to other computers that may not be on the same network.

Portable Storage Drives

There are two types of commonly used portable storage drives, USB flash drives and portable hard drives. A USB flash drive can store up to 16GB or more data files. It is convenient to use a plug-and-play USB flash drive. It has a size as small as a nail clipper and can be directly plugged into a USB port which is equipped on almost every computer. A portable hard drive can store up to 2000GB or more multimedia files. Many of these portable hard drives can use USB port. Unlike a USB flash drive, a portable hard drive often needs an external power supply and requires you to install software before you can use the hard drive, which is less convenient. When compared with a USB flash drive, the size of a portable hard drive is much larger and it is much heavier. Some of the large portable hard drives can have a weight over 10 pounds. A portable hard drive is good for storing movies and other multimedia files.

In the above, we have summarized some of the technologies commonly used by computer systems on the client side. The advance of technology is so fast that some of these technologies were not available just a few years ago and some of them may be out of date in the next few years. It is difficult to cover every technology that can be used for developing online computer labs. To keep up to date of the newest technologies, readers should keep track of the Web sites of major computer journals and IT product companies, such as

- http://www.cnet.com/
- http://www.pcmag.com/
- http://www.eweek.com/
- http://www.microsoft.com/
- http://www.cisco.com/
- http://www.redhat.com/

A discussion on how to select technologies for online teaching and learning will be given in later chapters.

Types of Web-Based Teaching Systems

Since online education entered the mainstream of education, software companies and e-learning consulting companies have developed hundreds of LMS and Web-based training management software packages. Various distance learning systems have been adopted for technology-based online courses. In this paragraph, we are going to have an overview of some of the Web-based teaching systems.

An early type of Web-based teaching system allowed an instructor to post text-based instruction notes on a Web site. It has some advantages over the traditional hand-outs. It is always available to students. Students are able to download these instruction notes anytime through the Internet so that they do not have to worry about misplacing or losing the instruction notes. This type of Web-based teaching system, however, lacks interactivity and multimedia course content. Since then, with the advance of technology, more sophisticated systems have been developed. The following are several commonly used Web-based teaching systems.

Partial Online System

In such a system, LMS is used to post course content online so that students can access the lecture notes, submit homework for grading, and take exams through the Internet. Once a week, for example, students will attend an instructor-led face-to-face lecture session where the instructor will go over the course content posted on the Web site and explain the projects to be completed in the computer lab. After the face-to-face lecture, the students will participate in the lab session for hands-on practice, since LMS does not support the lab for hands-on practice. After finishing the lab projects, the students can go back home to complete other assignments and upload the completed assignments to the Web site managed by LMS. The students can join the discussion group supported by LMS. The instructor grades the assignment and updates the grade report.

This system takes advantage of the face-to-face teaching platform, especially for the lab activities. It is much easier to set up an off-line lab. During the lab session, it is much easier for the students to engage in group activities. When an error occurs, it is easier for the instructor and students to work together to solve the problem. The disadvantage is that the students are not able to access the lab from anywhere and at anytime. If a student cannot attend the class on a specific day due to various reasons, he or she will miss the entire hands-on practice in the lab.

Online Lecture and Off-Line Lab System

This system takes lectures completely online. There is no face-to-face lecture session. Students can access the multimedia lecture notes posted on the LMS Web site at any time and from anywhere. The students can also submit their assignments and join discussion groups at the LMS Web site. The students are still required to attend lab sessions, say, every two weeks to complete their hands-on projects in an off-line lab. Although this system is a little more flexible than the partial online system, it does not really overcome the disadvantages.

WBT Lecture and CBT Lab System

In such a system, there is no off-line lab for hands-on activities. For the need of hands-on practice in an online technology-based course, currently, the most common solution is that students are required to have their own PCs at home and with the required software packages installed on their PCs. Publishers are aware of the requirement of hands-on practice by technology-based courses; the trial version of software sometimes comes with the book. Some of the online technology training companies also create multimedia CDs or DVDs to support computer-based learning. This way of supporting hands-on practice has its advantages and disadvantages. The greatest advantage is that students no longer depend on off-line computer labs on campus. They do not have to go to the campus for hands-on practice. They can work on their hands-on practice projects at home at anytime. This system is particularly convenient for courses that teach a programming language. Once a programming language software package is installed on a student's home computer, he or she can write code, debug the program, and upload the computation results to the LMS Web site for grading.

On the other hand, the system has some drawbacks. First, it is not a fully Web-based teaching system that students can access from anywhere. The software required for hands-on practice needs to be installed on students' computers. If a student has only a desktop computer at home, during traveling he/she will not be able to do hands-on practice. Also, although this type of system is good for a course that teaches a single IT product, it has difficulty when dealing with a course that involves hands-on practice on operating systems, networks, or on a client-server related structure. These hands-on practices often include multiple computers and networks. This kind of class may also require students to reinstall different operating systems. It is not a good idea to wipe out everything from the home computer to reinstall another operating system. Students may run into some hardware and software compatibility problems. Due to the difference among students' home computers, the installation and configuration of software may be different. Some students may have difficulty installing the software packages. The difference of students' home computers makes debugging more difficult. Without seeing a student's computer screen, instructors can only guess what has happened on the student side and students often misunderstand each other when they try to help their classmates.

As an alternative to CD/DVD which contains the software packages for a technology-based course, an LMS Web site may be configured to allow students to download the software for their hands-on practice. After downloading the software, students can send e-mail to the network administrator for the product key so that they can install the software on their home computers. In such a way, students can download the software anywhere and at anytime. There are some disadvantages for this method of getting software. It requires that students have a high-speed Internet connection due to the large size of the software package. Some of the software packages are so large that it may take several hours to download one package. Moreover, some of the software may get corrupted during the downloading process. After finding out that the installation files have problems, students have to spend more hours downloading the software again. Also, students need to have computers that are powerful enough to host these application software packages.

Complete Web-Based Teaching System

The rapid growth of the Internet technology has stimulated the development of more sophisticated Web-based teaching systems. WBT as an effective teaching and learning platform has been widely accepted by professional certification preparation institutions. Almost every professional certification assisting institution provides the choice of face-to-face training and Web-based training. WBT offers students flexibility and convenience in preparing for certification. Due to the high requirement for hands-on problem solving skills, many professional certification exams test students' hands-on skills. It is critical for students to learn hands-on skills before taking their professional certification exams. To get students ready for hands-on related questions on the exams, many professional certification preparation institutions have developed WBT with lab components. Through these types of WBT, students are able to access the simulated or real-life lab environment for hands-on practice.

Companies, such as Microsoft, Sun, and Red Hat, use WBT to provide training for their new products. Before even a new product is on the market, these companies post free e-training course materials to help potential users to get familiar with the product. Users will get a chance to learn about the new features provided by the forthcoming products. Often, these online training courses are accompanied with virtual labs that provide an in-depth, online hands-on training. Through virtual labs, users can learn how to install and configure new products. They can also learn how to use the new products to accomplish some of the administration and application tasks. These WBT courses are an effective way to learn about new products according to a user's own schedule.

The progress in developing Web-based learning tools and course management tools motivates colleges and universities to accept WBT as one of the teaching and learning platforms. Although Web-based training is popular in the IT industry, there are few reports from higher education institutions about this type of training for their lab-based courses. One of reasons is that it takes much more effort and resources to meet the requirements by the special needs of technology-based online courses. The second reason is the lack of prerequisite knowledge and skills. The trainees in the IT industry who participate in the WBT system are professionals. They are familiar with the fundamental knowledge and terminologies in the training of their own fields. Therefore, technology-based online training is not difficult for them. On the other hand, college and university students are new to the course content. They are not familiar with the basic terminologies used in the lab activities. So, they need more help from their instructors. Therefore, face-to-face teaching can do a much better job to help these students.

The biggest advantage of Web-based teaching is that it can reach out to the students who are unable to attend classes in a traditional classroom setting or who want an alternative to the classroom teaching. Online learning allows these students to control how and when they learn. No other teaching platform has this advantage. It is worth it to spend extra effort to develop Web-based teaching due to its flexibility and the fact that it can save money in the long run and increase enrollment for higher education institutions.

A complete technology-based WBT course may include four major components, a learning management system (LMS), live lecture and tutoring sessions, multimedia course content, and online computer labs. The description of LMS has been given earlier in this chapter. In the following, we will briefly discuss the other three major components.

Live Lecture and Tutoring

For live lecture and tutoring, class lectures are captured by Pan/Tilt/Zoom (PTZ) cameras. Students who participate in an online course can access the live video through a standard Web browser. The recorded live lecture can be used in two different types of WBT systems. To take advantage of the face-to-face teaching/learning platform, some of the higher education instructions require students to at least attend some of the live video broadcasts online at specific times so that the students have a chance to interact with the instructor and other students in real time. It is a live classroom environment without requiring students to travel to campus (Crane, 2006).

The second way of using captured lectures is to post them online. Students can view these lectures anywhere and at anytime. In such a way, there will be no online real-time interactivities with instructors and other students. The interaction can be performed through the discussion board and chat sessions.

Depending on the students' learning styles, some of them may prefer the real-time lecture so that they can discuss with the instructor and share their thoughts with other students. Other students may prefer to view the lecture online at their own pace. They can go over the content that is difficult to understand multiple times and they can review the course content any time when it is convenient for them.

An impressive live lecture tool is the virtual whiteboard. When a virtual whiteboard is connected to your desktop/notebook computer or a projector, it allows you to edit, write, or draw directly onto the screen. When it is used with a tablet PC, there is no keyboard or mouse to worry about since the pen acts as a mouse. A virtual whiteboard is good for labs and classrooms; it enables an instructor to share whiteboard notes with students in real time through the Internet. It also allows the students to electronically submit questions and comments. During the lecture, the instructor can use the virtual whiteboard to answer the students' questions. With the virtual whiteboard, the live class notes on the whiteboard can be viewed by students. Then, students can add their comments and save them for future review.

Similar to the live lecture, a tutoring session can also be live. Students can have live conversations with instructors through audio and video conferencing. Through audio conferencing, instructors can get feedback instantly. For most of the technical problem solving processes, it may take several rounds of testing to identify the problem and to come up with a solution. The instance information exchange is crucial for a problem solving process. Audio conferencing can be carried out through VoIP that uses the Internet's packet-switching capabilities to provide phone service.

Some of the Web conferencing software packages provide both video and audio conferencing tools. Video conferencing allows students and instructors to see and hear each other. In addition, it is also possible to let instructors and students to share computer screens. Web conferencing is a rich communication tool that offers new possibilities. It greatly improves the quality of the online teaching and tutoring. Video conferencing requires higher network throughput and faster computers on both the student side and the instructor side. Video conferencing can be transmitted over the Internet through three different technologies, ISDN-based video conferencing, IP-based video conferencing and streaming media. IP-Based video conferencing is less expensive although the quality of images is not as stable as the other two technologies. To have better quality, consider streaming media. Video conferencing is becoming part of the e-learning system to enhance communication, training, and instruction.

Multimedia Course Content

E-learning can be greatly improved by adding the multimedia course content such as graphics, animation, audio, and video. To overcome some of the disadvantages of e-learning, graphics play an important role. Graphics contain loads of information that are difficult to be described with words. In a WBT system, graphics may include charts, diagrams, icons, photographs, and drawings. Animations are great for demonstrating movements such as the moving of a mouse in multimedia lab instruction. When added to text-based lecture notes, voice, sound effect, and music can be used to explain complex concepts, and express emphases in lecture notes. Video content in online course materials can be used to improve the teaching and learning environment. It can make learning more active by communicating visually.

Online Computer Lab

Hands-on skills are crucial for a student to be a successful professional in technology-related fields. However, hands-on based courses have until recently been considered not suitable for Web-based teaching and learning. Unlike a WebCT-based online course which mainly provides online course content for students to access from client-side computers, for technology-based courses, students are often required to develop and maintain server-side projects, such as managing operating systems, building client-server computing architecture, developing enterprise-level database servers, and providing e-mail services. It is much harder to implement these types of projects through distance learning.

To allow students to perform hands-on practice on both the server and client sides, an online computer lab should have the following functionalities.

- Supports a broad range of operating systems, such as Linux and various versions of Windows operating systems.
- Allows students to create and manage various types of networks.
- Permits students to enforce security measures such as setting up firewalls.
- Allows students to practice special purpose server administrative duties.
- Is fully integrated with IT products on the current market.

In their short history, the earliest online labs for technology-related courses were created for electrical engineering programs. The following are some early online labs developed by several well known major universities (Fjeldly & Shur, 2003).

- **Automated Internet Measurement Lab (AIM-Lab):** It has been developed jointly by Rensselaer Polytechnic Institute (RPI) and the Norwegian University of Science and Technology (NTNU) since 1998. This online lab has been used for measurements of electronic devices through the Internet. The lab can be used by various senior undergraduate or beginning graduate courses in semiconductor devices and circuits. On the client side, through a Java applet, students can send commands to the server. On the server side, the commands are passed to the instrument drivers using the general-purpose instrument bus (GPRI) IEEE 488.2 standard protocol to operate the instruments. This lab allows instructors to monitor and control the server and modify the configuration of the instruments.

- **MIT Microelectronics WebLab:** It has been developed by MIT since 1998 for semiconductor devices and circuit characterization. The lab has been used for various electronic courses and by companies such as Compaq (now Hewlett-Packard Company) for electronic experiments. Some of the LMS functionalities such as online collaboration experiments, online tutoring, and assessment tools are also supported by WebLab. From a remote location, students can log on to WebLab through a Java applet which is the interface of WebLab. The Java applet checks the parameters entered by students. If the parameters entered by the students are correct, the students' requests along with the parameters will be sent to the IIS Web server. The requests are then passed to the driver VISA which converts each request to a set of GPIB commands that operate the computer-controlled electronic devices.

- **Next Generation Laboratory (NGL):** Again, this online lab is the joint effort of U.S.-Norway collaboration. It is designed for experimenting with analog integrated circuits. In this system, students can use the browsers on the client side to enter data into an online form and to view the graphics that represent the analog integrated circuits. A Web server is used to run the Web applications which handle the computation logic based on the requests from clients and support the online forms. The programming language C# is used to implement the computation logic. The second server is a lab server used to run Web services. Web services make the methods in the GPIB and Data Acquisition unit available to the C# program. Through the Web services, different lab setups can be connected to several other lab servers located at different geographical areas with no additional cost.

More information about the semiconductor online labs can be found on the AIM-lab Web (Shur & Fjeldly, 2007). This Web site gives detailed background information about the lab. Two sets of experiments, BJTs and Si MOSFETs, are provided for hands-on practice. The Web site also gives instructions on how to conduct the experiments online.

With these online labs, Web-based teaching is enhanced with hands-on practice on computer-controlled instruments. In the computer information systems curriculum, there are many courses that require hands-on practice on computers and network devices. Computers and network devices can also be considered as computer-controlled instruments. Therefore, the ideas on developing online labs for electrical engineering courses can be borrowed to create online labs for computer information systems courses. Indeed, many community colleges and universities are teaming up with IT product companies or computer training consulting companies to create Web-based training on a specific IT product such as Cisco network related training and training on Microsoft operating systems. The following are a few of the online computer labs developed by higher education institutions.

- **University of Texas at Austin:** By teaming up with the computer training company Element K, the University of Texas at Austin provides online computer training courses through its Thompson Conference Center. These courses are part of its continuing education program curriculum. Students can choose their courses from the online catalog and enroll online. The online courses offered at Thompson Conference Center are self-paced and Web-based. Students can take bundled courses in a particular subject area at their own convenience and pace within a 1-year period. Virtual labs are created to support hands-on practice for courses such as Cisco network and security related training, and Microsoft Certified Systems Engineer (MCSE). By practice in the virtual labs, students can learn the hands-on skills useful for developing real-life projects. Through a virtual lab, students who are taking the MCSE course can be connected to the actual Windows network, and students who are working on Cisco certification can operate on the actual Cisco routers and switches.

- **North Carolina State University:** North Carolina State University has been working on the Virtual Computing Lab project that is a collaborated effort of the College of Engineering Information Technology and Engineering Computer Services (ITECS) and the High Performance Computing (HPC) team in the Information Technology Division (ITD). IBM blade servers and other computers are used by the virtual lab. The virtual lab provides the Web services that allow remote access to computing resources. It also provides on-demand and reservation-based remote access to the extensive software library. The virtual lab is built to meet the increasing needs of both local and distant students and faculty for accessing the advanced computing laboratory facilities from anywhere and at anytime. The virtual lab eases the effort of updating and maintaining client computers. It allows any department to bring their computer labs online for their own students and allows students to access a specific computer lab from any remotely-accessible client computers.

- **Oklahoma State University:** Oklahoma State University plans to develop a virtual computer lab which is used to provide students with access to specialized computer lab resources from both on and off campus. The proposed virtual lab is based on technologies such as terminal services and VPN technology. Even though this is a virtual lab for student use from both off and on campus, it can be used to carry out hands-on practice for some technology-based courses. With the virtual lab, students are able to access the computer lab from any computer through an Internet connection. Students can choose operating systems among Windows, Macintosh, and Linux. They are also able to use the application software installed on the computers in the virtual lab and to use the network data storage device.

- **Red River College:** Red River College at Atlanta, Georgia developed a virtual computer lab running on a Citrix server. Citrix is a centralized special purpose server that delivers application software and data to client computers. It works like a VPN server that enables client computers to access Windows programs using the Microsoft Terminal Services software. The Citrix server based virtual lab expands the reach of the computer lab to students across town or even around the world. Students no longer need to install the application software on their own computers. This will save their money on software purchase and hard drive space. On the client computers, students need to install the Citrix plug-in before they can access Citrix. The virtual computer lab provides students and instructors with seamless, secure, on-demand access to computer resources. The Citrix based virtual computer lab allows computers with any operating system, Linux, Macintosh, or Windows, to access the application software through any types of Internet connections, including low-bandwidth and wireless connections. Another benefit of this type of virtual lab is that it provides strong protection to prevent virus and worm attacks of the server. Faculty members can also benefit from the virtual lab because this kind of virtual lab reduces faculty members' troubleshooting time and facilitates course development.

All of the above four systems showed that, by using the Internet technology, it is now possible to develop online computer labs that allow students to access the actual real-life computer lab environment for hands-on practice from a remote location. These online computer labs extend computer labs on campus to students' homes, hotel rooms, or offices for those nontraditional students. Through these online computer labs, students are able to access a university's computing resources 24 hours a day and 7 days a week. From a remote location, a student can directly operate the computers in a computer lab on campus just like he or she is sitting in front of a computer in a computer lab. By combining a LMS and an online computer lab, now we have adequate technologies to teach technology-based courses in which hands-on practice is an essential part of the courses.

For many universities, colleges, and community colleges, especially the small ones that lack funding and technical support, it is a challenging task to develop a Web-based computer teaching lab for technology-based courses. This book is designed to help readers overcome the difficulties in developing online computer labs. In later chapters, we will discuss the issues related to design strategies, implementation challenges, and the effectiveness of online labs.

Conclusion

In this chapter, we have gone over the technologies involved in developing online computer labs. The chapter has explained what these technologies are, why you might want them, and how to use them. It grouped the technologies into four categories, server technologies, learning management systems, network technologies, and workstation technologies. Servers play a crucial role in supporting the daily operations of an online computer lab. We briefly discussed the server operating systems and various special purpose servers. We also discussed the functionalities of LMS and the role played by LMS in the online teaching of technology-based courses. We looked into the functionalities of the network technologies and discussed their roles in an online computer lab.

For the workstation technologies, we discussed the issues related to audio and video hardware, software, and file formats. Various Media players, Web browsers, Web conferencing, and Web authoring software packages were also discussed in this chapter. In this chapter, we also compared the different types of Web-based teaching. We briefly discussed these Web-based teaching systems, and their strengths and weaknesses. To support the complete Web-based teaching system for a technology-based course, we often need an online computer lab. This chapter has given a brief review of some existing and future online computer labs.

The content covered in this chapter will prepare us to get familiar with the terminologies and equipment used in developing online computer labs in later chapters. Starting with the next chapter, we will get involved in the process of designing and implementing various online computer labs.

Since the future trends discussed in most of the chapters of this book are related, we will cover the future trends discussed in this chapter along with the future trends from other chapters in Chapter XII.

References

Correia, E., & Watson, R. (2006). VMware as a practical learning tool. In N. Sarkar (Ed.), *Tools for teaching computer networking and hardware concepts* (pp. 338-354). Hershey, PA: IGI Global.

Crane, E. (2006). Schools get virtual: New flexible collaboration tools can enable a better student-instructor distance learning experience. *University Business.* Retrieved April 1, 2007, from http://www.universitybusiness.com/page.cfm?p=839

Dara-Abrams, B. P. (2002). *Web technologies for multi-intelligent online learning.*

Retrieved April 1, 2007, from http://www.brainjolt.com/docs/webtech.pdf

Fjeldly, T. A., & Shur, M. S. (Eds.). (2003). *Lab on the Web: Running real electronics experiments via the Internet*. Hoboken, NJ: John Wiley & Sons.

Itmazi, J. A., & Megías, M. G. (2005). *Survey: Comparison and evaluation studies of learning content management systems*. Retrieved April 1, 2007, from http://scholar.google.com/url?sa=U&q=http://moodle.org/file.php/5/mod-data/forum

Kruse, K., & Keil, J. (2000). *Technology-based training: The art and science of design, development, and delivery*. San Francisco: Pfeiffer.

Lammle, T., & Tedder, W. D. (2003). *CCNA virtual lab, platinum edition (Exam 640-801)*. Alameda, CA: Sybex.

Lawson, S. (2005). *Windows leads as server market booms*. Retrieved April 1, 2007, from http://www.infoworld.com/article/05/11/23/HNwindowsleads_1.html

Minasi, M., Anderson, C., Beveridge, M., Callahan, C. A., & Justice, L. (2003). *Mastering Windows Server 2003*. Alameda, CA: Sybex.

Perry, T. L. (2000). *A history of interactive education and training*. Retrieved April 1, 2007, from http://www.refresher.com/!history2.html

Rowe, G. W., & Gregor, P. (1999). A computer based learning system for teaching computing: Implementation and evaluation. *Computers & Education, 33*(1), 65-76.

Sheltz, M., & Chellis, J. (2002). *MCSA: Windows 2000 virtual lab (CD-ROM)*. Alameda, CA: Sybex.

Shur, M. S., & Fjeldly, T. A. (2007). *Automated Internet measurement laboratory semiconductor device measurements using the Internet*. Retrieved April 1, 2007, from http://nina.ecse.rpi.edu/shur/remote

Valera, A., Díez, J. L., Vallés, M., & Albertos, P. (2005). Virtual and remote control laboratory development. *IEEE Control Systems Magazine, 25*(1), 35-39.

Veglis, A. (2000). Design of a Web based interactive computer lab course. In *Proceedings of the 10th Mediterranean Electrotechnical Conference: Vol. I* (pp. 302-305). Piscataway, NJ: IEEE.

Walton, M. (2005). Is "Wi-Fi on steroids" really the next big thing? *CNN*. Retrieved April 1, 2007, from http://www.cnn.com/2005/TECH/10/17/wireless.wimax/index.html

Wikipedia. (2007). *History of virtual learning environments*. Retrieved April 1, 2007, from http://www.answers.com/topic/history-of-virtual-learning-environments

Section II

Design of Online Computer Labs

Chapter III

Online Computer Lab Planning

Introduction

In Chapters I and II, we have done overviews about the issues related to online computer labs such as Web-based teaching (WBT) and various Web-based teaching systems. We have also briefly discussed technology-based courses and the technologies that can be used in the development of WBT and online computer labs. Starting from this chapter, we will discuss issues in designing and developing online computer labs for technology-based courses. The first task in designing online computer labs is to determine what type of online computer lab should be developed. It all depends on the teaching requirements. The developed computer lab should meet the needs of hands-on practice and balance the support, cost, and the complexity of technologies. This chapter provides you with a systematic way of identifying the needs of an online computer lab. In this chapter, we will discuss various approaches to decide what the teaching requirements are. We will start off with the topics related to online computer lab development process which will show you the big picture about developing online computer labs. Then, we will walk through the topics such as identifying hands-on requirements, identifying resources, and assessing costs. Investigating these topics will help you decide how big your project is, what the growth rate is, what the costs are, if there is funding for the project, and what kind of support you need to get from the computer service team. The next topic is about project planning which deals with issues such as budgeting, scheduling, forming a project development team, and implementing the project. This topic helps designers to deliver an efficient plan in developing a successful online computer lab.

Background

As mentioned in the previous chapters, the process of developing online computer labs is a complex process and requires careful planning. Even though it is hard to find publications that directly deal with the planning of an online computer lab, there are some publications that discuss each specific area of a planning process such as the planning of instructional design, Web page development, and the use of multimedia technologies.

In a planning process, the first task is to plan how to get the information from the key players who may impact the online lab development project. These key players may include university administrators, computer service department personnel, faculty members, and students. There should be a plan on how to meet these people and what information to collect. The information to be collected should include the assessment of the organization, budget, resources, and technical issues. Then, the collected information can be used to identify the challenges in developing online computer labs (Huntley, Mathieu, & Schell, 2005).

In the planning process, one needs to allocate the resources, form a development team, pick a WBT system, and build a framework for the project as pointed out by Horton (2000). The planning process should also include the preparation of teaching Web-based classes which will be supported by the online computer lab (Potter, 2003). To prepare the implementation and evaluation of WBT, checklists can be prepared for designing, developing, and implementing WBT as mentioned by Khan (2005).

Effectiveness is one of the concerns in the planning process. It is always a good strategy to create an effective Web-based teaching system with minimum resources. The issues of designing and developing effective Web-based training should be covered in the planning phase (Taran, 2003). The planning of effectively implementing Web-based teaching is another component of the planning process (Clark & Mayer, 2003).

One of the objectives of a planning process is how to effectively use technologies to improve instructional quality. Multimedia-based teaching materials are often used to achieve this goal (Clark & Mayer, 2003). As one of the components in a planning process, the planning of multimedia-based WBT is discussed by Lee and Owens (2004). One of planning tasks is to prepare useful tools and services for designing effective hands-on practice projects to support students in their learning. Some of interactive learning tools have been developed by Sarkar (2006) to assist the teaching of computer networking and hardware classes and to make teaching and learning more effective.

A planning process should help the online computer lab designer avoid pitfalls in designing and implementing stages (Khosrow-Pour, 2006). Some of the real-life case studies provided by Khosrow-Pour (2006) give successful and unsuccessful examples of the planning, design, maintenance, and management of telecommunications, network technologies, and applications.

In our lab development process, the online computer lab design consists of three phases, conceptual design, logical design, and physical design. The requirement analysis and planning stages belong to the conceptual design phase, where a designer can find out what the requirements are for the future online computer lab. In the logical design phase, the task is to translate the design objectives in technical terms. In the physical design phase, the task is to specify the technologies for the lab project. It is interesting to notice that the design model we use in an information system development process matches a similar model in the theory of instructional design. A generic model for designing teaching materials is called the ADDIE model which stands for analyze-design-develop-implement-evaluate. The online computer lab development process covered in this book basically fits into this model. This chapter discusses the stage of online computer lab planning. In the planning stage, lab designers will conduct analysis on the requirement for the online computer lab.

The idea of using a systematic approach to designing instruction is well discussed by Dick, Carey, and Carey (2004), who maintain that, in addition to the elements such as teachers, students, and teaching materials, the environment plays a crucial role in a learning system. There are many other instructional design approaches and models. The book edited by Reigeluth (1999) summarizes various theories and models in the instructional design and celebrates diversity in the field of instructional design. For readers who are interested in the application of instructional design in the fields of online learning pedagogy, computer conferencing, and electronic collaboration, they may find some great ideas from the book edited by Bonk and King (1998).

Building an online computer lab is creating a learning environment and an online computer lab that promotes hands-on practice and interactivity can greatly benefit from the research of instructional design. Readers who want to learn more about the trends of instructional design theories and practice may consider Reiser and Dempsey's (2006) book which discusses the trends and issues in the instructional design field and how technologies will affect the future of this field.

In this chapter, a more systematic approach is discussed for planning online computer lab projects. First, we will have an overview of the stages of an online computer lab development process which provides clues for what should be planned. We will then identify the requirements for teaching and hands-on practice. We will also identify the requirements for resources such as budget, equipment, labor, technical support, and so on. The next important planning stage is the cost assessment in which you will analyze the cost of hardware and software, consulting, labor, utilities, knowledge update, and so on. In the project planning stage, the project manager should draft a timeline and form a team for the project. The investigation of network architecture, Internet service providers, and other technology related subjects should be part of the planning. The planning of training should also be part of the planning process. The planning should address the issues related to lab evaluation and upgrade. A task list should be prepared for each stage of the project. Funding for the project should be allocated. The project manager should also prepare the agreements and contracts for the project so that all the people involved in the project are clear about their duties.

Online Computer Lab Development Process

As mentioned in Chapter I, online computer labs are necessary for some of the online technology-based courses. In addition to learning management systems (LMS), we need to develop special online computer labs. Unlike general purpose computer labs that are designed for surfing the Internet and hosting application software such as Microsoft Office and some programming language packages, the online computer labs used to support technology-based online courses are constructed on a client-server structure. To perform some system and network related tasks, students have to have an administrator's privilege. In Chapter I, it was emphasized that we need careful planning before we can implement a computer lab like that. In the following, we will go through some major steps in a lab development process.

Requirement Analysis

The first stage in the design of an online computer lab is to identify requirements by a technology-based course. Before an online computer lab can be built, we must collect information about the objectives to be achieved by the future computer lab. A well-designed online computer lab is based on a good understanding of the needs of the hands-on practice. At this stage, one or a team of lab designers will conduct interviews with the instructors and the technical support team to find out:

- What technology-based courses require an online computer lab?
- What are the required technology skills for performing hands-on practice?
- How many students will simultaneously log on to the online computer lab?
- What are the hands-on activities?

Based on the results of the interviews, we can set up the design objectives for the computer lab to accomplish and the sequence of activities to be performed. The Sequence Diagram in Figure 3-1 is a quick example to illustrate how to interpret the needs of remote access to an online computer lab.

With the objectives and the activity sequence, you can verify if the online computer lab will meet your targeted lab teaching requirements. The process of developing a computer lab will not only deal with technology issues, but also need to deal with other important issues such as budget limitation, technical support, and product comparison. It is a good idea to develop a project plan to include all the issues for consideration.

Figure 3-1. Sequence diagram for online computer lab access

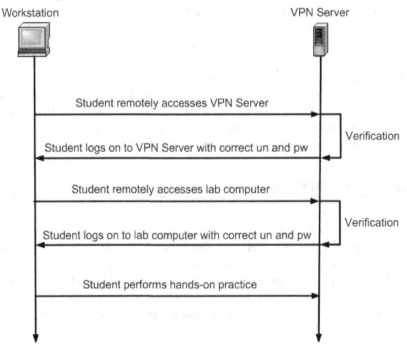

Project Planning

Often, a computer lab that satisfies all the requirements can be a complicated system. Therefore, a careful plan is necessary for the success of building the lab. For a big project, you may need to document the conclusions drawn from the interviews. In the document, the users' views about the project should be listed. For more details, you may want to include some case studies to illustrate how exactly students perform their operations. Budgeting and purchasing are also part of the planning. You need to answer questions such as:

- What is the estimated cost for developing and maintaining the online computer lab?
- Is the funding available?
- Do we have the technical support personnel to handle the lab development and management job?

At this stage, you will first organize a lab development team, select software and the related computers and network equipment. It is the designers' responsibility to determine the type and model of the equipment. As a project manager, you need to draft a timeline to illustrate the journey that accomplishes the design goals. The last step of the planning stage is to get approval from the decision maker. Once the online computer lab plan is approved, you will enter the lab design stage.

Lab Design

A solution design includes three phases, the conceptual design which presents the users' views about the project, the logical design which presents the designer's view about the project with the technical terminologies such as objects and interaction, and the physical design which selects the physical components for the project. In our lab development process, the requirement analysis and planning stages belong to the conceptual design phase. Once you have figured out what the requirements are for the future online computer lab, the next step is to translate the design objectives in technical terms and select the technologies for the lab project. This stage belongs to the logical design and physical design phases. At this stage, as a designer, you should be able to answer questions such as:

- What are the requirements for the computer system that will be used as a server?
- How can multiple courses share the same lab resources?

- What is the software to be installed on the server and clients?
- How can the network be constructed?
- What are the requirements for the network equipment?
- How can the students remotely access the online computer lab?
- What are the specifications of the computers and network devices?
- How can the security be set up to prevent malfunction from damaging the server and network meanwhile allowing students to have an administrator's privilege?
- How are student accounts managed?

To get the whole picture of the client-server structure on which the online computer lab will be developed, you may want to model the entire system with a flow chart and a network topology map. For each node on the flow chart, detailed information can be further specified. After the flow chart is created, you can use it to verify if the future online computer lab will meet the teaching requirements. Usually, several modifications are needed. It might take some time to make sure that the technical support team and instructors both understand and are satisfied with the lab design.

- **Lab implementation:** During the implementation process, the lab designer, instructors, and the technical support team should work together to get the job done. The technical support team will construct the network and install the hardware and software. For some small colleges and universities, instructors and designers may also be involved in the installation of hardware and software. For a large university, an online computer lab may be physically located in multiple locations and hundreds of students may log on to the online lab simultaneously. This will require teamwork by the people from different campuses.

- **Lab testing:** After the hardware and software are installed, it is the time to test the newly created online computer lab before it can be used for online teaching and hands-on practice. Based on the task flow chart, you need to test the entire system including the activities on both the server side and client side. Make sure that you will be able to access the server from the client side. The testing results should be recorded for future reference. Based on the testing results, the lab development team may need to further modify the lab configuration in order to meet the teaching requirements.

- **Developing lab teaching materials:** Once the online computer lab is ready to use, instructors can create lab teaching materials for hands-on practice. The lab teaching materials may include the detailed lab manual, multimedia demonstration files, and some troubleshooting instructions. Through the Internet, an instructor may use a student account to test the lab teaching materials from a remote location.

- **Online computer lab management:** It is the technical support team's job to keep the online computer lab running smoothly. At small colleges and universities, faculty members are often involved in the lab management and maintenance. The management of a distributed online computer lab system can be a challenging task. Most students do not have experience with the server-side operations. However, they are practicing to be system administrators. It is easy for these students to make vital mistakes and cause system crashes. The daily maintenance tasks may include system backup and recovery, performance tuning, troubleshooting, technical support for students and instructors, enforcement of security measures, and students' account management. In fact, the most common help needed by students is solving their log-on problems.

Final Thoughts on the Online Computer Lab Development Process

The field of information systems is changing rapidly. As a result, textbooks and lab teaching materials are updated accordingly. For each semester, there will be some new requirements for online computer labs. This will restart the lab development cycle again, from the requirement analysis stage to the lab management stage. The technical support team and instructors need to work closely to update the online computer labs before a semester starts. The lab update may also require budgeting and purchasing of new hardware and software.

In an online computer lab development cycle, the analysis of requirements is the first stage. The issues related to the requirement analysis will be discussed next.

Identifying Hands-On Practice Requirements

To answer questions such as what the hands-on practice requirements are, you need to thoroughly understand the teaching materials. In this session, we are going to discuss how to identify the hands-on practice requirements. Several commonly used methods and procedures will be discussed here.

At the beginning of this identification process, the lab designer needs to make appointments to interview instructors. The designer should list all the questions before the interviews. The following is some of the information the designer should collect:

- The objectives of the hands-on practice.
- The hardware and software currently used by the hands-on practice.
- The requirements for students' accounts.
- The activities to be performed by the students.
- The maximum number of students who will simultaneously log on to the same server.
- The type and amount of data generated by the hands-on practice.
- The requirements of collaboration among instructors and students.
- The requirements for system performance.
- The requirements of multimedia lab teaching materials.
- The difficulties of using the current system for hands-on practice.

The following are some suggestions that may be useful for collecting the information.

- You may organize a meeting which includes administrators, instructors, and the technical support team to initialize the communication and set up long term goals.
- When you contact instructors to make interview appointments, you may want to e-mail the instructors your questionnaire for them to prepare for the interviews.
- During interviews, make sure to get all the answers for your questionnaire. Based on the feedback from the instructors, ask some new questions and get clarification from the instructors.
- If necessary, take a closer look at the lab assignments and projects that need to be done by the students. If these documents are related to your design project, make sure to examine them carefully. You can also observe how the students work on their hands-on practice in the current system. Ask about how these assignments are done currently and how the future online lab can make a difference.
- Good interpersonal and communication skills are greatly helpful for collecting information.
- By the end of each semester, e-mail instructors to ask about possible updates for the online computer lab for the upcoming semester, or set up a feedback Web page for the instructors to submit the new update requirements.
- Carefully review the collected information and identify the design objectives. Document the conclusion drawn from the collected information.

Once the information is collected, you need to categorize it and analyze the requirements for the online computer lab and interpret it in technical terms. Depending on the requirements, the next task is to identify the resources which will be discussed next.

Identifying Resources

After the hands-on practice requirements are identified, the lab designer's next task is to identify the resources to support the online computer lab project. Depending on the needs of hands-on practice, the online computer lab can be a very complex project that needs various resources to support the development process. Before you can start the development process, you need to gather all the available resources to see if the underline project is doable. The resources to support the development of an online computer lab may include four major components: administrative support, skill and knowledge update, technical support, and financial support.

Administrative Support

The support from administrators is critical for the success of the online lab development project. As mentioned in Chapter I, among all the support needed for developing an online computer lab, the administrative support is the most important. It is difficult to get financial support and technical support without the support from administrators. The following are some possible supports we can get from administrators.

- It is very important to encourage instructors to be involved in the development process. To motivate and encourage the instructors, administrators can organize meetings with them to emphasize the importance of WBT and identify the needs of the instructors. The administrators can provide stipends and course release to the instructors who participate in the development process. They can also make the instructors' effort be counted towards their tenure and promotion.

- It is equally important to motivate the computer service department to be involved in the development of the online computer lab. The administrators should organize meetings between instructors and technicians so that they can understand each other better.

- Approval of additional funding to hire additional technicians to support the online computer lab is another great thing the administrators can do.

- The administrators can also help with advertising the online lab project to the local community by writing articles for local newspapers and television stations. They can pass the information to local community leaders to seek additional support.

Skill and Knowledge Update

The rapid progress of technology requires students, instructors, and the technical support team to constantly update their skills and knowledge to keep up with the current trend. Possible resources to update skills and knowledge are:

- **Workshops:** There are many professional development training companies that provide training of new technologies. For those who cannot participate in face-to-face lectures, these training companies will also provide Web-based training. The advantage of workshops is that you get to talk to people with a lot of firsthand knowledge about the new technologies. The disadvantage is that the workshops are designed for professionals, so it may not be suitable for less experienced trainees. Often, in these workshops, the trainers end up talking about their work experience instead of showing you step-by-step how to get the job done.

- **Internal training seminars:** Many colleges and universities offer short training seminars to instructors and students. At the beginning of each semester, students who are enrolled in online courses are to take an online course orientation seminar. They can take the seminar either online or face-to-face. In the seminar, the students will learn how to log on to the course Web sites, how to submit their homework, how to take quizzes and exams, how to collaborate with other students and communicate with instructors. Short and focused training seminars offered to instructors will demonstrate how to use LMS to manage a class, how to post teaching materials, and how to use multimedia tools. The online teaching support team can even provide one-on-one consultation for instructors. Most of these training seminars focus on using a general purpose LMS. For the development and usage of online computer labs, instructors and the technical support team have to get additional training.

- **Professional conferences:** Almost every faculty member is a member of one or more professional associations in their own fields. Each year, these associations organize various conferences for faculty members, students, administrators, and e-learning practitioners to present their papers. In these conferences, you will learn new strategies, theories, results of analyses from your colleagues.

- **Books:** Books are another great resource for updating your skills and knowledge. You can choose the book that meets your expectation and learn the content according to your own pace. The disadvantage of books is that you may not fully understand the content and there may not be someone to answer your questions for clarification. Some small colleges and universities have limited funding for libraries. These libraries, especially, do not like to purchase books for technology professional development since these books will be out of date soon and there is little value to keep them in the libraries.

- **Web sites:** This is one of the best resources to update your knowledge. Major technology product companies always keep their Web sites updated. When running into technical problems, you can often find solutions posted on these Web sites. The only drawback is that the information on the Internet is not always organized systematically. It often takes a great amount of time to find what you want.

Technical Support

It is almost impossible to develop and manage an online computer lab without technical support. Technical support is ranked the second most important resource according to Pirani's (2004) survey. Usually, the technical support team is able to provide general support on using LMS. For the special need of an online computer lab, the technical support team tends to shy away from it due the following reasons.

- The technologies used in online computer labs are different from those that technicians usually manage. The security measures are quite different, too. The differences complicate the technical support.
- It is so easy for students to mess up with the machines in the computer lab trying to learn how to be system administrators or database administrators; this makes the technical supporting team's job more difficult.
- Online computer labs need to be updated every semester. The updating process takes too much time.
- It is difficult to answer students' questions since their questions are often related to the course content which a technician may not be familiar with.

Understanding the teaching requirements is an important step for the technical support team to be motivated to support online computer labs. From the top level, the administrator needs to promote the conversation between the people involved in teaching and the technicians. The teaching team should understand the limitation of the computer and network resources and the concern of security problems. On the other hand, the technical support team should understand that teaching is the revenue generating division. It is the technical support team's duty to support teaching. If the quality of teaching causes the drop of enrollment, the technical support team will be downsized too. The administrators, instructors, and technicians should work on an agreement on the support issues before the lab development gets started.

Financial Support

Developing online computer labs needs financial support for the costs of hardware, software, network equipment, technical support personnel, and many other things. Financial support for technology is one of the most important resources for developing online computer labs. However, for many colleges and universities, this is also a very challenging issue to deal with. Especially at some small campuses, the budget for computer labs is very limited. Online computer labs for teaching technology-based courses need to support both the server-side and client-side computation. This type of lab costs more on equipment and technical support. Before you can find a solution to overcome financial obstacles, it is difficult to start the online computer lab project. The following are some possible solutions.

- **Educate the decision makers about the importance of the lab project:** You should develop a budget proposal with the objectives of the project, needs analysis, benefits of the project, comments from students, and case studies of similar projects from other colleges or universities.

- **Identify companies and organizations that may provide grants for lab construction and upgrade:** If there is an opportunity, write a proposal to these companies and organizations. A university's president usually can help with fundraising.

- **Propose a budget jointly with the computer service department:** Many administrators may not understand the needs of online computer labs. The computer service department often has a bigger influence on administrators in this area. Also, there may be a proposal writer in the computer service department. The person may have the experience to better present the needs of online computer labs.

- **Charge some additional fees for online technology-based courses:** Use this method cautiously. It is not the best way to solve a financial problem.

- **If the funding is not available to purchase new computer systems, you may consider using surplus PCs on campus as a backup plan:** Each year, some computers in the offices and general purpose computer labs are replaced by newer computers. Some of those replaced computers may be adequate to support most of the hands-on practice for the technology-based courses. For each online lab, the cost of network equipment can also be reduced by configuring the surplus PCs as routers which are the key network equipment.

- **Use open-source software such as Java, MySQL, and Linux:** The use of open source software can reduce cost. On the other hand, it takes more time to learn different brand products and takes a longer time to make these products work together. Different brand products also make technical support more difficult.

The effort to gain financial support takes a lot of energy, but it is necessary and rewarding. With the financial support, the online computer lab developing project can now take off.

Assessing Costs

Based on the hands-on requirements, the designer will first estimate the costs of the project. The following are some possible costs for an online computer lab.

- **Cost of consulting:** For some large universities, an online computer lab may be built by a consulting company. In such a case, the cost should include the charge for the initial lab set up and the continuous support of running the lab. Usually, you will let multiple consulting companies bid on the project, and use one that best fits with your design objectives.

- **Cost of furniture:** If it is a new lab, you should consider the cost of furniture such as desks, chairs, network connection ports, and electrical facilities. You may also include the cost of room remodeling.

- **Cost of computer hardware and network equipment:** The designer can search the Web sites for prices of servers and network equipment. Several Web sites provide product information and user comments about the products. Often, these Web sites provide reports on the evaluation of the products. Sometimes, they also provide side-by-side comparisons of similar products. Some of the popular Web sites for product information are:
  - http://www.cnet.com/
  - http://www.pcmag.com/
  - http://www.pricegrabber.com/
  - http://www.bizrate.com/
  - http://www.techbargains.com/

  Other costs may include memory upgrade, adding new hard drives, printer and paper, network cables, storage devices, and multimedia equipment.

- **Cost of software:** For higher education institutions, the cost of software may not be a big concern. Many software companies offer an education price which is much lower than the market price. Some of the major software companies support higher education institutions with special academic programs. For example, Microsoft offers the MSDN Academic Alliance (MSDNAA) program (MSDN Academic Alliance, 2005) and Oracle offers the Oracle Academic Initiative (OAI) program (Oracle Academic Initiative, 2002).

- **The cost of MSDNAA is about $800 per year:** The MSDNAA program allows many of the Microsoft software products to be used for education purposes. Under the MSDNAA contract, the software products included can be used by an entire department. Each faculty member or student in the same department can install the software on a desktop computer or notebook computer at home or in the office. Most of the server-side software such as Windows Server 2003, SQL Server 2005, and Microsoft BizTalk Server are included in the program. This program also includes some client-side software such as Windows XP Professional and Microsoft Access. Many application development software packages such as Visual Studio .NET, which contains the programming language software C++, C#, J#, ADO.NET, VB.NET, ASP. NET, are also provided by the MSDNAA program. The OAI program is another great program for supporting the online computer lab development. The cost of the OAI program is also a few hundred dollars per year. OAI provides database related software for educational use. The products include database server software, application server software, database application development software, and data analysis software. Sun Microsystems (2005) provides the open source Java packages which include the Java application development software such as J2SE and J2EE. You can download the Java products from the Sun Web site free. For programming related courses, The J2SE and J2EE packages can also be installed in the lab for courses requiring Java programming. Another useful open source software product is Linux, such as Red Hat Linux. The Linux operating system has been used in networking and system administration courses (Petersen & Haddad, 2004). Red Hat charges a small service fee for the academic edition of Red Hat Workstation and Red Hat Enterprise Linux (Red Hat Academic Products, 2005).

- **Cost of technical support:** For a large online computer lab, someone from the computer service department needs to be assigned to the computer lab for daily maintenance and technical support and someone needs to take care of the budgeting and purchasing. Some online computer labs also hire student workers. The cost of the technicians and student workers should also be included in the annual budget.

- **Cost of knowledge update:** The cost of training can be a significant portion of the total cost. To let students learn the most updated knowledge, instructors and technical support team members should be encouraged to update their own knowledge by participating in workshops, training classes, and conferences. Funding should be allocated to support these activities.

Project Planning

As mentioned before, developing an online computer lab is a complex process. It is worth it to take some effort to work out a feasible project plan. If we simply start the project without planning carefully and later find out it does not meet the online hands-on practice requirements, it can cause a lot of chaos to the online computer lab development process. Many classes may depend on this lab. In such a case, it is almost impossible to rebuild an online computer lab all over again during the middle of a semester. To successfully construct an online computer lab, you need a plan that may include the following components:

- Objectives of the online computer lab.
- People involved in the project, including administrators, instructors and support staff, and duties preformed by these people.
- Consulting services.
- Required hardware and software.
- Room and building information.
- Network architecture and equipment.
- Internet service provider.

Figure 3-2. Online computer lab objectives

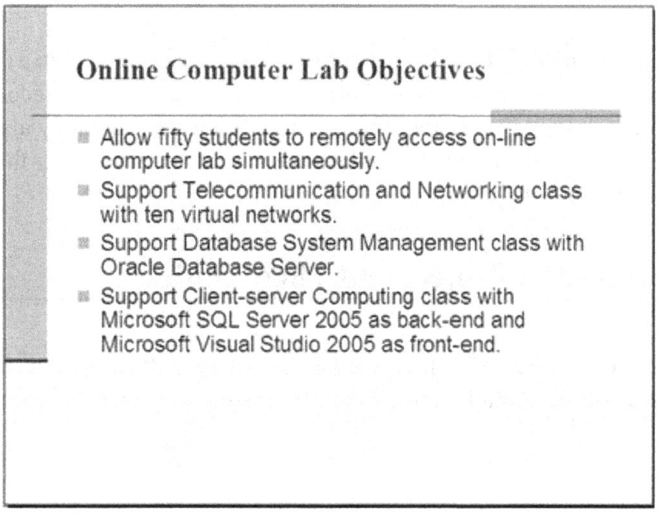

Online Computer Lab Objectives

- Allow fifty students to remotely access on-line computer lab simultaneously.
- Support Telecommunication and Networking class with ten virtual networks.
- Support Database System Management class with Oracle Database Server.
- Support Client-server Computing class with Microsoft SQL Server 2005 as back-end and Microsoft Visual Studio 2005 as front-end.

- Security measures to be enforced on the online computer lab.
- Technical support issues.
- Training arrangement.
- Task list.
- Timeline for the entire project
- Review and evaluation process.
- Online lab maintenance and upgrade.
- Budget preparation.

Each online computer lab project is different. The components above are for general purposes. A specific plan may include more or less of those components. In the following, let us examine each component in detail.

Objectives

Objectives give a guideline on what to achieve. The objectives of an online computer lab should match the requirements of teaching and hands-on practice. They should be measurable and achievable. For example, you may draft some objectives shown in Figure 3-2.

Be specific about what to achieve. In that way, you will have a clear idea about who should be involved in the project and what equipment should be purchased.

Consulting Services

A college or university may contract the entire or part of their project to a consulting company. The contract may include information about tasks to be accomplished, payment, project timeline, and other information. Some contract may also include information such as the charge for troubleshooting and upgrade. After the contract is signed, keep it in a safe place.

People Involved in Project and Their Duties

To achieve the objectives, you first need to organize a development team. The team members should include administrators, instructors, and technical support personnel.

- **Administrators:** It is the best to include the department head or college dean and the director of the computer service department in the team. When a conflict occurs, the leaders can work together to find a solution.

- **Instructors:** Instructors' involvement is crucial. They are the ones who are responsible for developing lab teaching materials and can provide useful information about the requirements of the online computer lab. Instructors may also be involved in the online lab development and maintenance. Make sure to include the instructors of the courses that require the online computer lab.

- **Technicians:** Technicians will be responsible for developing the online lab. They will also provide technical support for students and instructors. Due to the fact that the development of the online computer lab requires various skills, you should include the network manager, database administrator, security manager, application developer, help-desk manager, e-mail manager, Web page developer, and technical supporter in your team. At a small college or university, some of these roles can be combined together.

- **Student workers:** If you have student workers who assist in the lab development, include them too.

- **Other support personnel:** You may also include secretaries and people who will take care of ordering products and managing the finance.

The above people will be involved in every stage of the development process. You will need assistance from these people during the development of the online computer lab. A member of the team needs to understand the other members' work and closely work together with them.

Hardware and Software

The planning of an online computer lab deals with the requirements for hardware and software. Document all the requirements for hardware and software. In the document, describe the usage of the hardware and software and how they can be used to meet the requirements. You may also want to include all other information about the hardware and software, such as the vendor contact information, models, prices, and product technical specifications. Sometimes, your proposal may not get fully funded. To prepare for this, you should rank the priority of the hardware and software. It is not necessary to purchase everything for the online computer lab, some of the hardware and software may already exist in your college or university. You may check the computer service department's inventory list and include the information about existing hardware and software that can be used for the online computer lab in your plan. The selection of hardware and software will also depend on the performance requirements. We will discuss this topic later in the physical design phase.

Figure 3-3. Online computer lab network map

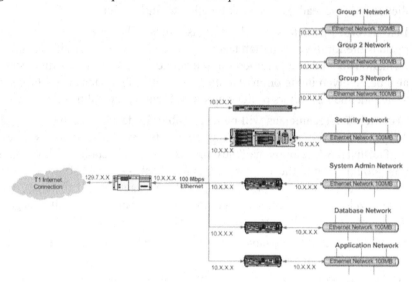

Room and Building Information

Dedicate a room to keep the computers and network equipment. The room should be secure and equipped with furniture, air conditioning and fire alarm facilities. The power supply should be adequate for running the computers and network equipment. To prevent the damage of power outage, an uninterruptible power supply (UPS) device should be properly installed. If the room is shared with others, you will need a storage room or cabinet to keep the computer lab documents, parts, spare computers, and network equipment.

Network Architecture and Equipment

The network architecture and the detailed configuration information of the online computer lab should be well documented for future reference. Create a table that contains each computer's host name, IP address, subnet mask, and location. You may also do so for each router in your network. For each subnet, write down its network ID and number of computers connected to the subnet. If a directory service is built for the network in the lab, you should record the information about the domains, sites, DNS servers, DHCP servers, security settings, and group policies. For a more sophisticated computer lab, network printers and network data storage devices may also be installed. Include the information about these devices in your plan document, too. Figure 3-3 shows an example of network architecture.

Internet Service Providers

For the Internet connection, it is important to write down the Internet service provider (ISP)'s contact information. You will need the information to get technical support from the ISP. Make a note about the information provided by the ISP such as the Internet server's IP address, DNS server's IP address, and the IP address for the gateway. Record the protocols currently running on the Internet router. You may need to enable a certain protocol for a specific remote access method.

Security Measures

Enforcing security measures will protect the online computer lab from being damaged by unauthorized users. However, it may also cause some problems for remote access and prevent students from performing certain activities. There is often a need to release some of the security measures to allow students to do hands-on practice. Knowing what security measures are currently enforced will allow us to make correct decisions on modifying the security measures. The following are some of the security measures that should be useful for an online computer lab.

- Enable services and open ports based on a list of roles played by the students and the requirements from the client side.
- Restrict access to some specific ports and configure some ports so that the content through the ports can be digitally signed or encrypted.
- Configure communication protocols to prevent password cracking and man-in-the-middle attacks.
- Configure Web server security to prevent anonymous users from accessing content files.
- Configure auditing based on the auditing policies which can be specified as: Not audit any events, audit only successful events, or audit both successful and unsuccessful events.

Technical Support Issues

There should be an agreement on what technical support the computer service department can provide. Unlike the general technical support which commonly involves providing help on desktop application related issues such as e-mail problems, the technical support for the online computer lab involves solving server-side and network problems shown below.

- Students may need help related to technology-based course content such as configuring the Web server for running ASP.NET programs.

- Technicians may also have to deal with security conflicts. A lot of hands-on practice in technology-based courses requires users to have the administrator's privilege and that is a headache for technical support.

- Students can easily make mistakes while configuring a new system. It is much harder to find and correct a mistake in a misconfigured system file that causes the system to crash.

- The technical support team also needs to deal with problems on different computing platforms such as Linux based systems and Windows based systems.

Technical support of online computer labs needs experienced network administrators, system administrators, database administrators, and technicians to help on these issues. However, these types of technical personnel are highly needed by other departments, too. They may not have time to focus their attention on what happens in the online labs. Instructors may help with some of these issues. To efficiently manage work orders from students and instructors, the computer service department can create a database to support a work order form which should be posted on the department's Web site. When a work order comes in, it will be saved in the database. Based on the nature of the service request, the computer service department can arrange a technician to do the work requested by the work order. Figure 3-4 is a sample Microsoft Access work order form created with a template provided by Microsoft Office.

An agreement should be written on which part of technical support is the computer service department's responsibility and which part is the responsibility of instructors.

Training Arrangement

Identify the skills needed to achieve the goals of the project. If a certain skill is lacking for the development of the online computer lab, you need to identify the type of training and funding for the training. Some of the training can be done internally, some can be done through Web-based training, and some may require face-to-face instruction. Document the training needs and the training schedule.

Figure 3-4. Sample work order form

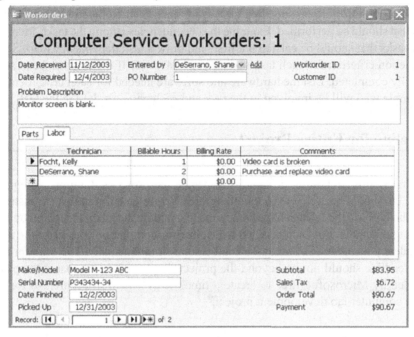

Figure 3-5 shows a possible training schedule which is based on a template provided by Microsoft Office.

Task List

Figure 3-5. Training needs

Job Position	Department	Required Training	Training Title	Description	Availability	Provided by (Organization)
Network Manager	Computer Service	Network security	Cisco Network Security Solutions	Protect online computer lab network and reduce network operating costs	External	Network Training
Server Technician	Compter Service	Advanced server skills	64-Bit Computing Migration	Migrate current 32-bit server to next generation 64-bit server	External	Server Training
Desktop Technician	Computer Service	Troubleshooting of application software	Technical Support for Dell Desktop Computers	Help users solve technical prolems of application software	Internal	Computer Service Department
Database Administrator	Computer Service	Setting up database clusters	Oracle 10g and Real Application Clusters	Improve written and business communication skills	External	Database Training
Instructor	Informaton Systems Department	Using live Web conferencing software	How to Use Breeze	Help coworkers get up speed on technical tasks and processes	Internal	Computer Service Department

Training Needs Source Data / Course Titles per Position / Cours

Based on the information specified above, for each of the objectives, list all the tasks that should be performed. Examine the dependencies among the tasks, identify those tasks that should be carried out first, and draw a task dependency map. Set a completion criterion for each task so that you can verify if the task has been successfully completed. List the hardware and software needed for each task. Identify the people who will be involved in the task and each person's duty in the task.

Timeline for Entire Project

The first step in creating a timeline is to decide the project's duration. Often, an online lab should be competed a month before a semester starts so that the instructors who will use the lab can have some time to evaluate the lab and develop online teaching materials. For each task, set a deadline for completion. Based on the task dependency map, some of the tasks can be done simultaneously. The time for the final deadline should not be beyond the project duration. You can use a template provided by Microsoft Office to create a timeline as shown in Figure 3-6 for an online computer lab development project.

Review and Evaluation Process

Figure 3-6. Online computer lab development timeline

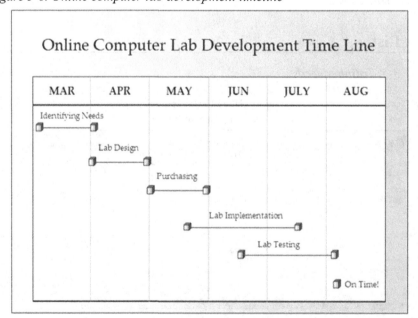

After the online computer lab is completed, ask the instructors to give it a test by running some lab activities to see if the objectives of the project are achieved. It is important to realize that the lab development process is improved gradually with feedback from users. You may also let student workers test the lab from their homes to see if the remote access is configured properly. After testing, you often need to make some modifications on the current project. Make sure that these modifications are done promptly since it is close to the beginning of a new semester and there is little time to fix additional problems.

Online Lab Maintenance and Upgrade

To keep the online computer lab running smoothly, you need to develop a maintenance plan which includes the maintenance tasks and the person(s) responsible for the tasks. The maintenance plan should include a system backup schedule and measures for disaster recovery. The online computer lab needs to be upgraded for each new semester due to the following reasons.

- New courses may be offered in the new semester,
- New textbooks may be adopted,
- Technologies in the lab need to be upgraded, and
- The lab may get additional funding for adding new hardware and software.

In the plan, you may need to include the activities to be performed for upgrading the lab and the people in charge of the upgrade. Figure 3-7 created with a Microsoft Office template shows a list of possible maintenance tasks.

Budget Preparation

Create a spreadsheet that contains detailed information about the estimated cost of the online computer lab project. In the spreadsheet, list the following projected costs.

- Payment for consulting services and other labor.
- Expense of hardware and software and network equipment.
- You may also need to include the cost of the Internet connection.
- The cost of lab maintenance and upgrade.

Figure 3-7. Computer lab maintenance tasks

	A	B
1	**Computer Lab Maintenance Schedule**	
3	**Daily**	
4	**Desktop Computers**	
5	Software Update	Check web sites for update patches for update.
6	Computer Systems	Check if computers are turned on. They should be on 24/7 for remote access.
7	Security	Check if computers are infected by virus and fix problems if yes.
8	Multimedia Equipments	Check if multimedia equipments are working properly.
11	**Network**	
12	Network	Test network to see if computers in the lab are able to communicate.
13	Security	Monitor network to see if there is unexpected attack to network.
14	Internet	Check if Internet is properly connected to allow remote access.
18	**Server**	
19	Event log	Inspect event log to see if there are error messages and fix problems.
20	Backup	Verify is system backup is working properly.
21	Computer Systems	Check if system is working properly after power outage.
22	Software	Check if application software is running properly.
23	Security	Update patches or service packages if they are available.
26	**Room**	
27	Air Conditioning	Check if air conditioning works properly.
28	Power Supply	Check power supply. If there is power outage, call for help.
30	**Every Semester**	
31	**Lab**	
32	Lab Construction	Modify or reconstruct lab architecture based on teaching requirements.
33	Upgrade	Replace old computers with newer ones in each three-year cycle.
34	Installation	Install new hardware and software for same classes or for different classes.
36	**Hardware**	
37	Repair	Repair computer hardware problems.
38	Equipments	Add new multimedia equipments to existing computers.
39	Upgrade	If necessary, add additional hard drives and more memory.
44	**Software**	
45	Reconfiguration	Reconfigure software for different classes.
46	Repair	Fix or reinstall software crashed by students.
49	**Network**	
50	Reconfiguration	Reconfigure network for different classes.
51	Repair	Repair or replace misconfigured network equipments.
52	Upgrade	Add new network equipments to existing network.

Computer Lab Maintenance

- The cost of furniture and room remodeling.

To cover some unpredicted cost, a small amount of reasonable overestimate is helpful.

All the costs can be divided into the cost for purchasing and the cost for operations. The purchasing cost can be divided into two categories, the category of new equipment and the category of surplus equipment that needs repair and upgrade. In the budget, you should specify the quantity of each item and the subtotal for each category. You also need to include the information of the vendors from whom you will purchase the products. Some states give a list of vendors from whom a state

Figure 3-8. Online computer lab budget

Online Computer Lab Development Budget

Organization: Jackson University   Year: 2008

Department: Information Systems and Computer   Submitted by: Jay Smith

Annual allotment: $ 200,000.00

Equipment: $100,000   Maintenance: $20,000   Total Budget: $180,000

Consulting: $50,000   Other: $10,000

Equipment and Software Budget

Line	Item	Vendor Information	Qty.	Unit Cost/Rate	Total
1	ABC S-0016 Server	www.abc.com. 800-567-1234. Jay Smith	2	$ 16,000.00	$32,000
2	AnyWhere LMS	www.AnyWhere.com. 800-123-1111. Mark Fry	1	$ 20,000.00	20,000
3	ABD D-14 Desktop System	www.abc.com. 800-567-1234. Jay Smith	15	$ 1,000.00	15,000
4	R-0011 Router	www.abc.com. 800-567-1234. Jay Smith	1	$ 7,000.00	7,000
5	SW-1000 Switch	www.abc.com. 800-567-1234. Jay Smith	5	$ 200.00	1,000
6	U-1090 UPS	www.DependableUPS.com. 800-001-1111. Ed Taylar	1	$ 9,000.00	9,000
7	Db Softwre	www.db.com. 800-111-1234. Angie Ada	1	$ 1,000.00	1,000
8	Os Software	www.os.com. 800-111-5401. Jen Stone	1	$ 1,000.00	1,000
9	Security Software	www.security.com. 800-450-9000. James May	1	$ 1,000.00	1,000
10	Lab Management Software	www.manage.com. 888-123-5555. Juan Rodriguez	1	$ 1,000.00	1,000
11	Multimedia Software	www.multimedia.com. 877-111-1111. Ana Diaz	1	$ 2,000.00	2,000
12	San-100 Network Storage	www.san.com. 866-578-1087. David Chen	1	$ 10,000.00	10,000
13					0
				Grand Total	$100,000

Consulting Service Budget

Line	Item	Vendor Information	Qty.	Unit Cost/Rate	Total
16	Security Consulting	www.secure.com. 800-110-9000. Linda Lee	1	$ 9,000.00	9,000
17	System Consulting	www.system.com. 866-530-0000. Dave Thomas	1	$ 18,000.00	18,000
18	Network Consulting	www.network.com. 888-450-1113. Mary Garza	1	$ 12,000.00	12,000
19	Database Consulting	www.database.com. 800-111-5560. Larry Jung	1	$ 11,000.00	11,000
				Grand Total	$50,000

Maintenance Budget

Line	Item	Vendor Information	Qty.	Unit Cost/Rate	Total
30	Security Upgrade	www.secure.com. 800-110-9000. Linda Lee	1	$ 500.00	500
31	Technical Support	Computer Service Dept. 570-4220. Joe King	1	$ 10,000.00	10,000
32	Hardware Purchase		1	$ 2,000.00	2,000
33	Software Update		1	$ 1,500.00	1,500
34	Employee Training		6	$ 1,000.00	6,000
35					0
				Grand Total	$20,000

Other Budget

Line	Item	Description/Justification	Qty.	Unit Cost/Rate	Total
36	Travel		6	$ 1,000.00	6,000
37	Room Remodeling		1	$ 4,000.00	4,000
				Grand Total	$10,000

college or university can purchase products. In such a case, make sure that the vendors you use are on the list. If the budget is limited, you need to check the current surplus inventory to see if you can use any of those surplus computers and network equipment. The purchasing cost can be considered an annually-based budget. It should be budgeted once a year.

For the cost of operations, you may include the cost for security update, memory upgrade, hard drive upgrade, software upgrade, and so on. The estimated cost for replacing broken parts should also be included. If a network printer is installed in the lab, you need to include the cost for the toner and paper. In some project, student workers are hired for technical support. If so, add the budget for the cost of student workers. For security reasons, a university may let the online computer lab use a separate Internet connection. If that is the case, include the cost of the Internet service in your budget. If there is a demand for live audio and video for the multimedia based teaching materials, the cost for VoIP and the connection that allows video and audio streaming should also be included in the budget. The cost of operations can be event-based. In Figure 3-8, a possible budget for an online computer lab is given.

What you have budgeted may not be approved due to various reasons. Often, you may have to think about a backup plan. You may have to use an alternative way to achieve the objectives with lower expense.

Conclusion

This chapter has dealt with the issues related to online computer lab planning. First, we took a look at each stage of the development process. As the first stage of the online lab development process, we explored various ways of collecting information. This chapter has also demonstrated how to interpret the collected information in technical terms. We have investigated the resources for building an online computer lab. We have discussed the support of administrators, professional development, technical support, and issues related to funding for the development of the online computer lab.

Next, we talked about the cost for building an online computer lab. We have explored some alternative ways to reduce the cost so that our spending will not exceed the limited funding. The last topic covered in this chapter is the planning of an online computer lab development. This chapter has given detailed descriptions about how to identify the objectives of the project, prepare the document for consulting service, gather information about hardware, software, room, and network structure, and draft an agreement on technical support. We have also looked into training for technicians, instructors, and related personnel to gain the skills for developing and managing the online computer lab.

Lastly, an example of a detailed budget plan was given in this chapter. Sample charts, tables, and graphs were included to illustrate some components of an online computer lab development plan.

The In the next chapter, we will consider some strategies for developing various types of online computer labs based on the hands-on requirements of technology-based courses.

References

Bonk, C. J, & King, K. S. (Eds.). (1998). *Electronic collaborators: Learner centered technologies for literacy, apprenticeship, and discourse.* Mahwah, NJ: Lawrence Erlbaum Associates.

Clark, R. C., & Mayer, R. E. (2003). *E-learning and the science of instruction: Proven guidelines for consumers and designers of multimedia learning.* San Francisco: Pfeiffer.

Dick, W. O., Carey, L., & Carey, J. O. (2004). *Systematic design of instruction* (6th ed.). Upper Saddle River, NJ: Allyn & Bacon.

Horton, W. (2000). *Designing Web-based training: How to teach anyone anything anywhere anytime.* New York: Wiley.

Huntley, C., Mathieu, R. G., & Schell, G. P. (2005). An initial assessment of remote access computer laboratories for IS education: A multiple case study. *Journal of Information Systems Education, 15*(4), 397-407.

Khan, B. H. (2005). *Managing e-learning strategies: Design, delivery, implementation and evaluation.* Hershey, PA: Information Science Publishing.

Khosrow-Pour, M. (2006). *Cases on telecommunications and networking.* Hershey, PA: Information Science Publishing.

Lee, W. W., & Owens, D. L. (2004). *Multimedia-based instructional design: Computer-based training; Web-based training; distance broadcast training; performance-based solutions* (2nd ed.). San Francisco: Pfeiffer.

MSDN Academic Alliance. (2005). MSDN Academic Alliance Developer Center. Retrieved January 12, 2005, from http://www.msdnaa.net/

Oracle Academic Initiative. (2002). Oracle Academic Initiative. Retrieved January 11, 2005, from http://oai.oracle.com/en/index.html

Petersen, R., & Haddad, I. (2004). *The complete reference: Red Hat Enterprise Linux & Fedora edition.* Emeryville, CA: McGraw-Hill/Osborne.

Pirani, J. A. (2004). Supporting e-learning in higher education. ECAR. Retrieved April 26, 2006, from http://www.educause.edu/ir/library/pdf/ERS0303/ecm0303.pdf

Potter, T. (2003, April). Configuring compute labs: Training needs dictate computer lab design. *AALL Spectrum Magazine,* pp. 16-19.

Red Hat Academic Products. (2005). Red Hat Academic Products. Retrieved January 10, 2005, from http://www.redhat.com/solutions/industries/education/products/index.html#WS

Reigeluth, C. M. (Ed.). (1999). *Instructional-design theories and models: A new paradigm of instructional theory, Volume 2.* Mahwah, NJ: Lawrence Erlbaum Associates.

Reiser, R. A., & Dempsey, J. V. (Eds.). (2006). *Trends and issues in instructional design and technology* (2nd ed.). Upper Saddle River, NJ: Prentice Hall.

Sarkar, N. (2006). *Tools for teaching computer networking and hardware concepts.* Hershey, PA: Information Science Publishing.

Sun Microsystems. (2005). The network is the computer. Retrieved January 11, 2005, from http://www.sun.com

Taran, C. E. (2003). *Standalone WBT.* Timisoara: Orizonturi Universitare Publishing House.

Chapter IV

Strategies for Developing Online Computer Labs

Introduction

In Chapter III, we have discussed the issues of planning for an online computer lab. From the collected information, the online lab designer can get the users' views about the project. The next task is to design a lab that will meet the design objectives. To accomplish this task, you need to first develop a model of the online computer lab. Then, use the model to verify if the objectives have been achieved. Modeling allows you to select appropriate architecture for the lab project. During the modeling process, you need to illustrate the flow of activities. We will discuss the modeling issues in this chapter. We will first investigate the types of lab architectures and analyze how lab architecture fits a specific design objective. For each of the architectures, we will discuss the strategies to select technologies to be used to construct the lab.

After the model is developed, we will deal with issues in the physical design phase. We will specify the hardware and software requirements for various technology-based courses. Different courses may also require different network structures. In this chapter, we will look into some network design related topics. After we have walked through the topics about the selection of hardware, software, and network equipment, we will consider the issues related to remote access. It is another important topic to be covered in this chapter. This chapter provides various possible remote access plans and the selection of a remote access schema.

Background

A lab system development process can be considered a special case of a general solution development process. A solution development process contains two components, software development and infrastructure deployment. The development of a lab system partially involves both components on a smaller scale. Solution design processes can be applied to various projects such as modeling a health system (Chong, Clark, Morris, & Welsh, 2005) and design of a grid computing system (Meliksetian et al., 2004).

Modeling service-oriented solutions is a process to convert the user's view about a project into a logical model which will be used for the implementation of the project (Erl, 2005). In a modeling process, the server-side architecture needs to be specified. The designer has to make a decision on the selection of the server-side architecture among many possible choices (Chevance, 2005). The decision should be based on the project objectives and the challenges to be faced in the project (Wiehler, 2004). In this chapter, we will summarize the strategies on meeting the challenges for the development of online computer labs.

Major IT product companies such as IBM, Microsoft, and Cisco provide service to clients who implement the systems with their products. Solution design is a tool used by these companies to implement the systems for their customers. Therefore, these IT companies often require that solution design be a subject for professional certification. IBM has been using solution design for globalized architecture (Stearns, Zhu, Cui, Shu, Xu, Li, Li, & Qu, 2004). The globalized architecture allows the execution of a program to be processed across multilingual data and to be presented in a culturally correct way. To be certified, one needs to pass the exam, IBM Certified for On Demand Business-Solution Designer (Ransom et al., 2005). The Microsoft certification program offers the Microsoft Certified Architect (MCA) program for solution architecture (Microsoft, 2006). The program provides the latest information about solution architecture and modeling tools (Andersen, McDermott, Dial, & Cummings, 2005). The training materials are available to help IT professionals to be MCA Microsoft certified architects (ePlanetLabs, 2005). The Cisco Solutions Architecture Fundamentals exam is about the Internet solution architecture (Recore,

Laurenson, & Herrmann, 2003). It is used to test IT professionals on a variety of feasible solutions to network infrastructures.

To implement a logically designed solution, one needs to specify the hardware and software to be used in the solution. This chapter will cover some topics on hardware and software selection. The implementation process includes the selection of server hardware and software (Marcraft International, 2005), network equipment (Groth & Skandier, 2005), and remote access technologies such as virtual private network (VPN) (Bollapragada, Khalid, & Wainner, 2005).

Some of the ideas, methods, and procedures in a solution development process can be borrowed for developing our online computer lab project. A solution development process may contain five models, business architecture model, team model, infrastructure model, process model, and application model. In Chapter III, the topics such as identifying the requirements for hands-on practice, identifying resources, assessing costs, and project planning are all parts of the solution development framework.

Although the author is unable to find previous research on the application of the instructional design theory directly to the computer lab project, there are some research studies that have been done on technology training. Lewis, Wimberg, Kinuthia, Guest, and Jackson (2003) applied the instructional design theory to the faculty development workshop which provides training on developing online learning environments. Another interesting study by Guo, Katz, and Maitland (2002) reports the evaluation of the instructional design applied to an online anatomy lab course at the University of Calgary. Their study shows that, when planning the online course, it is important to set clear course objectives and to know the students' academic levels, learning styles, and what motivates them. A lot of research has also been done in the area of course content development. Chapter IX will provide a brief review on studies in this area. Next, we will further look into the issues on the design of an online computer lab from the logical perspective.

Strategies of Technology Selection

On selecting technology products, it is a bit confusing since there are many similar products on the market for you to choose. To prepare to purchase hardware, software, and network equipment for the online computer lab, you have to make decisions on selecting the right service and products. The following are some strategies that can be used to help you make decisions.

- **Decide who will operate lab:** You will first need to decide if the online computer lab will be built and managed in-house or outsourced. Due to the concern of costs, the needs of technical support, and high frequent upgrading and redesigning, most online computer labs should be built and managed in-house.

- **Decide lab type:** You also need to decide what type of online computer lab will meet the teaching and hands-on practice requirements. Later, we will give descriptions of small, medium, and large online computer labs to help you make a decision.

- **Compare similar products:** Next, you will need to investigate the functions and features of a product and then compare these features with other similar products. We will discuss how to select hardware for implementing the lab infrastructure and how to select software for solving a business problem which, in our case, is the hands-on practice required by various technology-based courses.

- **Choose vendors:** You need to decide if the products will be purchased from a single vendor or multiple vendors, and the pros and cons for a computing platform. It is ideal to get all the products from a single vendor. However, it is often true that a single vendor does not have all the products you need and a product from a specific vendor may not be as good as that from other vendors. On the other hand, if the products are from multiple vendors, you need to consider the compatibility of these products and if there are problems to get services from these vendors.

- **Decide how servers will be used:** To properly choose a server, we first need to know how the server will be used in the lab. You will need to find out if the server will be used for basic network management, for hosting application services such as e-mail and databases, for supporting multimedia collaboration, or for managing a grid computing environment.

- **Decide how servers can be accessed:** It is important to know who will be using the computers in the lab and how they can remotely access the server. You need to investigate the network technologies used to connect the clients and servers and make a decision.

If there is no best solution, as a decision maker, you need to consider a trade-off on costs, features, convenience, compatibility, and support.

Online Computer Lab Models

To create an appropriate architecture for the online computer lab, you are required to first develop an architecture model for the online computer lab project and then use

Figure 4-1. Peer-to-peer lab model

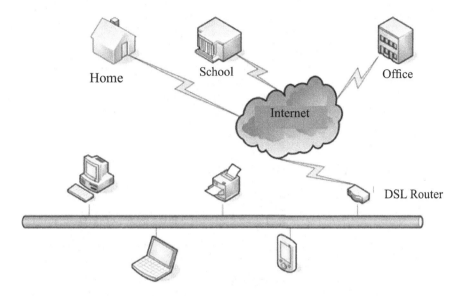

the model against the objectives of the computer lab. The model is used to represent the infrastructure of the online computer lab and will be used as a blueprint for lab development. If things do not work out as originally expected, revise the model to meet the design goals.

Based on the class enrollment and the technologies to be used for the teaching and hands-on practice, the online computer lab can be constructed in several different ways. Let us first consider some lab infrastructures in the following.

- **Peer-to-peer lab model:** This lab model is designed for small classes with the enrollment of less than 10 students and the classes only need client software for hands-on practice. The peer-to-peer solution provides an environment for collaboration among computers. The computers in the lab can communicate through a local network and share some services such as Internet connection, printing, file sharing, and accessing services with other computers by using the Remote Desktop Connection utility. Note that there is no server involved in this solution. This type of solution is easy to set up and does not require server-side hardware and software, and advanced network equipment. It is suitable for small programming related classes or application development classes. A model for a peer-to-peer lab solution is given in Figure 4-1.

- **Small lab model:** This model is designed for classes with less than 30 students. For the small lab model, one or more servers are configured to provide services such as system backup and recovery, directory service for user and computer

Figure 4-2. Small lab model

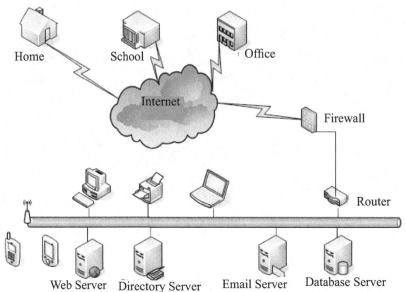

management, firewalls, e-mail messaging, local and remote client networking, file sharing, Web services, and printing service. This kind of lab is relatively easy to set up and manage. It provides a secure computing environment. With the support of a variety of services, the small lab model is suitable for system administration courses, networking courses, database system courses, e-commerce courses, Web development courses, and security management courses. It is also a low-cost solution for most medium-sized classes. A model for a small lab solution is given in Figure 4-2.

- **Medium lab model:** This model is designed to support several different types of technology-based courses simultaneously. It provides an integrated infrastructure that has the flexibility and scalability to meet the teaching and hands-on practice requirements of a small- to medium-sized computer science or information systems department. Multiple servers and network structures are constructed to support a variety of courses. This model also supports the services such as proxy server technology, system backup and recovery, directory service for user and computer management, e-mail messaging, local and remote client networking, file sharing, Web services, and printing service. The medium lab model will cost more, especially for the servers and network equipment. It will take a longer time to build and need more skills for server development and network construction. Figure 4-3 is a model for a medium lab solution.

- **Large lab model:** The large lab model is designed to support technology-based courses offered by a medium and large computer science or information systems

Figure 4-3. Medium lab model

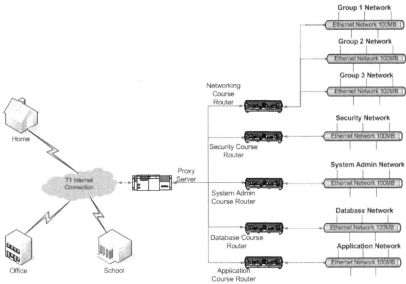

Figure 4-4. Large lab model

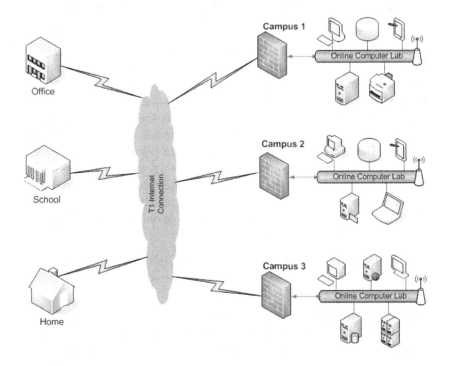

department. The lab system can be distributed across multiple campuses that may be located globally in different countries. Being a distributed system, it has great reliability and scalability. It can avoid the single point failure and is available 24 hours a day, 7 days a week for teaching hands-on practice. It may be constructed on a multitier architecture. The large lab model may include several structures similar to those for the medium lab model. In addition, it may also include enterprise-level application software such as data warehouse software, middleware messaging software, and mobile networking services across heterogeneous platforms. The construction of a large computer lab takes a great deal of effort and knowledge. Usually, it is done by the joint effort of a large university and an IT consulting company. Figure 4-4 demonstrates a large lab model.

Depending on the requirements by the technology-based courses and the scale of the campus, you can choose one of the models for your online computer lab project. Most of the computer science and information systems departments should be able to handle small or medium-sized online computer labs.

Online Computer Lab Solution Design

After a lab model is selected, the next task is to design solutions for teaching and hands-on practice. A solution design process investigates a business problem and then translates it into a series of technical terms so that the business problem can be solved by using technologies. It is crucial that the solution designer well understands the requirements. An efficient method to make the hardware and software work together to the meet the requirements is to use a flowchart to represent the lab activities. The flowchart contains the objects that represent the hardware and software on which the students will carry out hands-on practice. As you go through the flowchart, you will be able to verify if the requirements are satisfied. You will also be able to identify the hardware and software needed for teaching and hands-on practice. The following are the tasks in developing an online lab flowchart.

- Identify the objects in the model. The objects in a lab model are hardware equipment, software packages, and network structure of the lab, and user interface for remote access.
- Identify the lab regulations, such as only the students who are enrolled in the class can access the online computer lab, or each student can only see the database project developed by the owner of the database, and so on.

- Define the standard for the user interface, terminologies, and technologies to be used in the lab. A unified standard will make the development more consistent. It can also make the communication among the developers and the users easier.

- Create the model with graphical tools.

- Validate the model to make sure that the design goal has been achieved.

- If the model is not completely satisfactory, modify the original model and test again.

In addition to the objects, a model should also include the activities shown below.

- **Remotely access lab computers:** Users access the computers in the lab from remote locations with correct user names and passwords assigned by the computer service department.

- **Log on to lab computers for class:** Users access the computers in the lab with user names and passwords assigned by an instructor for a specific class.

- **Use programming software:** Use various programming languages to implement business logics and develop an application interface.

- **Use application software:** Develop applications such as forms and Web pages with application software. Students can also use application software for multimedia presentations and can submit their homework.

- **Manage operating systems:** Use system management software to deploy distributed computing structures, administrate user accounts, and develop directory services.

- **Construct network:** Construct enterprise-level network structures. The activities may include creating subnets, configuring network equipment, monitoring the network activities, and sharing files.

- **Develop database systems:** Use database management system software to create, manage, and query databases.

- **Build client-server architecture:** In the lab, students create networks that support client-server computing. They also create the back-end, middle tier, and front-end with various software packages. With the client-server architecture, students can implement an e-commerce system by developing storage devices, Web servers, and databases.

- **Perform analysis:** Students can carry out performance analysis for operating systems, networks, and databases. They can also conduct analysis related to business intelligence such as data mining.

- **Configure security measures:** Students learn to enforce network security measures and encrypt sensitive data.

Figure 4-5. Online computer lab flowchart

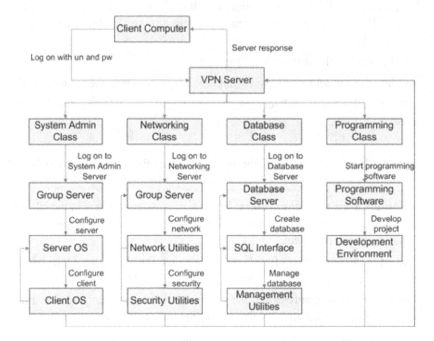

Figure 4-5 shows a simplified flowchart of an online computer lab that supports teaching and hands-on practice for the technology-based courses: System Administration, Networking, Database Management, and various programming courses. Each rectangle in the model represents an object in a lab infrastructure. The activities are marked along each direction line. The flowchart is for illustration purposes only. You can add many more details to the flowchart, or you can use another flowchart for the details. To develop a more detailed flowchart, readers can refer to some solution design reference books.

The model starts from a client computer at a remote location where a student is trying to access the online computer lab. With the user name and password assigned by the computer service department, the student can log on to a VPN server which is one of the computers attached to the university's network domain. The VPN server is protected by the intense security measures enforced by the computer service department. After logging on to the VPN server, the student will log on to the server configured for each individual class, or log on to one of the computers on which the programming software packages are installed. All the students need their user names and passwords assigned by the instructors to log on to the computers dedicated for individual classes. For some classes such as the Networking class, the students are further divided into groups. Each group of students can log on to their server with their user names and passwords for that group. For the classes such as System

Admin and Networking, students must be given a system administrator's privilege. After logging on to the computers for each class or each group, the students can perform hands-on activities.

 The above two paragraphs have covered some important modeling issues for the design of an online computer lab. We have discussed lab infrastructures by examining various lab models. We have also discussed some solution design issues. Next, we will investigate the strategies on the selection of hardware and software.

Selection of Hardware

Hardware related technologies may include computer systems, network equipment, data storage equipment, printing equipment, and image creation and display equipment. In this section, as part of the physical design process, we will discuss the selection of hardware related to computer systems and network equipment.

Computer Systems

There are two types of computer systems used in an online computer lab system, server and client computer systems. The top four server suppliers are IBM, Hewlett-Packard Company (HP), Sun, and Dell. Servers can be constructed on a stand-alone minitower computer, a rack frame which can hold multiple servers, or a blade enclosure which can also hold multiple servers but takes up much less space. For the purpose of online computer labs, entry-level or midrange-level servers should be adequate for most cases. To select appropriate servers for online computer labs, let us first look at the characteristics of each type of server. An online computer lab can use Intel or AMD-based servers or use UNIX-based servers which vary from one vendor to another. The following types of servers can be used to meet the needs of an online computer lab. Based on their structures, the servers can be categorized as stand-alone, rack system, and blade system.

- **Stand-alone server:** A stand-alone server may have one or more processors. It may have up to 64GB or even more RAM to support a large number of simultaneous operations. It can support multiple internal hard drives with the capacity of storing 4TB or even more data. A stand-alone server is often equipped with input/output devices such as one or more monitors, floppy disk drives, CDRW/DVD drives, tape backup systems, or other portable storage devices. It often has multiple network cards to reduce network bottlenecks.

- **Rack system:** Usually, an online computer lab may have multiple servers running at the same time. To efficiently manage these servers, a rack system is a good choice. A rack system is a collection of servers interconnected in a rack frame. For each server on the rack, one or more processors can be installed. 12GB or more RAM can be installed on each server. The servers can have their own hard drives for data storage or share a network storage device such as a Storage Area Network (SAN) device for the highly scalable data-intensive Web learning system. A SAN device can also be mounted on the rack frame. A SAN device has over 1GB sustained network throughput and it can store over 120TB data. Another advantage of SAN is that it is compatible with various operating systems such as Microsoft Windows, Sun Solaris, HP-UX, IBM AIX, and Linux from different distributions. Similar to a stand-alone server, each server in a rack system can have its own input/output devices.

- **Blade system:** A blade system is an alternative of a rack system. With a common backplane, it is much easier to manage blade servers. A blade system needs fewer interconnection network cables, less electricity, and fewer KVM monitor switching devices. A blade system has the ultimate in scalable performance and capacity. It takes much smaller space than a rack system. Its dense enclosure can hold 60 or more blade servers with each powered by single-core and multicore processors. To make a blade system highly available, blade servers are hot pluggable. It may also be equipped with multiple power supplies, cooling fans, I/O modules, and hard drives.

A server can be built by using a 32-bit or 64-bit microprocessor. The use of a 64-bit system allows the server to handle more memory and larger files. With much better performance, 64-bit processors are also able to handle the applications that require high computing power such as audio/video processing, numerical modeling, gaming, querying a massive database. The following is some of the advantages of a 64-bit system.

- **Performance:** If 64-bit native software is available, the performance of a 64-bit system is much better. A 64-bit CPU can handle as twice as much data per clock cycle when compared with a 32-bit system.

- **Memory:** The amount of memory that can be allocated by a 32-bit system is 4GB. With the 64-bit system, in theory, you can allocate 18 quintillion bytes. In practice, it can allocate over one terabyte (TB) or more memory.

- **Flexibility:** A 64-bit system may also be able to run 32-bit software without sacrificing the performance.

Table 4-1. Server platforms

Platform	Manufacturer	O/S	Description
x86	Intel or AMD	Windows, Linux, UNIX, Solaris	An x86 server is designed around one or more 32-bit or 64-bit single-core or multicore x86 microprocessors.
POWER	IBM	AIX, MacOS,	A POWER server is constructed based on POWER chips. 32-bit or 64-bit applications run equally well on the 64-bit architecture. A POWER server can serve up to 4,906 or more processors and memory up to 2 terabytes.
SPARC	Sun	Solaris, Linux	A SPARC server is constructed on 64-bit UltraSPARC or SPARC64 processors. The server can run 128 or more processors simultaneously. It is a highly scalable and reliable server. It can also address over 500GB memory.
Itanium 2	HP/Intel	Windows, Linux, HP-UX, z/OS, OS390	Itanium 2 is a 64-bit RISC microprocessor. Itanium can run on the 32-bit mode. However, it is much slower when running at the 32-bit mode. Itanium 2 can support up to 512 processors running simultaneously and allocate 60 TB memory.

Servers can be built on various platforms. The descriptions of different platforms are listed Table 4-1.

As seen in Table 4-1, SPARC, POWER, and Itanium are primarily designed for large applications. For a computer lab project, the x86 platform is often adequate except for a large lab model. For a small lab model, one or more stand-alone servers are usually adequate. For a medium lab model, a group of stand-alone servers, a rack system, or a blade system should be good enough. For a large lab model, a combination of stand-alone servers, rack systems, and blade systems are often used to handle the distributed computation.

It takes great effort to build and run a reliable online computer lab. The selection of the right devices requires a lot of experience and knowledge. The selection of hardware devices is also limited by the following factors.

- **Reliability:** A server should support a stable computing environment. It is important that the server is highly dependable during each semester. The measure of reliability is the server's uptime. The server is in an excellent condition if it has 99% or better uptime. When a problem occurs, the server should be fixed as quickly as possible.

- **Maintainability:** The measure for maintainability is the average time spent on each server in each semester. The maintenance of servers often takes a lot of time and effort. When managing and upgrading the servers during a semester, you have little time left to get things done. A server that needs less maintenance is a desired factor for server selection.

- **Deployment:** For a computer lab, it is desire that a server can be quickly deployed for classes. It is important to choose a server that is easy to reconstruct and upgrade.

- **Support:** The selection of a server also depends on what type of support it can get from the computer service department. Often, the computer service department is only familiar with certain types of servers. You need to work with the computer service department to see what types of servers it can support.

- **Ease of use:** The time and effort spent on how to operate a server is another concern for selecting servers.

- **Data storage management:** A large online computer lab may require servers to interact with network data storage equipment such as Storage Area Network (SAN) devices. The selected servers should be supported by the SAN devices.

- **Cost:** There are two kinds of costs, initial cost and operation cost. The initial cost may include the cost of hardware, software, consulting, and room renovation. The operation cost may include the cost of maintenance, upgrade, salaries, and training. Often, there is a limit on the budget. The total cost cannot be more than the budget.

On the client side, the configuration of a desktop or notebook computer may vary. It will depend on what kind of computer a student may own. In addition to the basic configuration which includes CPU, RAM, Hard drive, input/out devices, CDRW/DVD, and a monitor, a client computer should also be equipped with a sound card for multimedia course materials. It should also include a microphone and Web camera for online conferences. Most of the personal computers sold on the current market satisfy these requirements.

Uninterruptible Power Supply (UPS)

To prevent servers from being damaged by an unexpected power outage, each server system should be supported by an uninterruptible power supply (UPS). An UPS system uses generators and/or batteries to power itself. During a power outage, the UPS system can keep operating and preventing the server from losing services for mission critical applications. The selection of an UPS depends on the devices to be protected by the UPS. The following are some devices to keep in mind when you select an UPS.

- **Desktop computer:** For a minitower desktop computer with Pentium IV CPU, 17in CRT monitor, ATA100 IDE hard drive, CD/DVD drive, and support of 20 minute holding time during a power outage, you need a UPS that can support 500VA power.

- **Stand-alone server:** For a midrange stand-alone server with two Inter Xeon CPUs, a 19 inch CRT monitor, two ultra 320 SCSI internal hard drives and one external hard drive, one CDRW/DVD, switch or router, and support of 20 minutes holding time during a power outage, you need a UPS that can support 1000VA power.

- **Rack system:** For a midrange rack system with four servers, each of which has one Intel Xeon CPU, one ultra 320 SCSI internal hard drives, one CDRW/DVD, switch and router, and support of 30 minute holding time during a power outage, you need an UPS that can support 3000VA power.

- **Blade system:** For a midrange blade system with eight servers, each of which has one Intel Xeon CPU, one ultra 320 SCSI internal hard drives, one CDRW/DVD, switch and router, and support of 30 minute holding time during a power outage, you need an UPS that can support 12000VA power.

Making a correct decision on the selection of a UPS system is important for running a successful online computer lab.

Network Equipment

The possible network equipment used by an online computer lab are network cards, switches, routers, Internet connection service equipment, and network storage devices. Routers are the main network equipment which are used to connect various networks. There are different types of routers based on their functionalities. In Table 4-2, we compare some of the routers that can be used to construct online computer labs.

Switches are also important network equipment. They are used to connect other network equipment. The switches in Table 4-3 can be used for the construction of online computer labs.

Another kind of useful network device is the Storage Area Network (SAN) device optional to medium or large computer labs. SAN connects servers of various platforms to a centralized pool of data storage devices such as hard disk array and tape devices. SAN can significantly reduce the operating system backup time and

Table 4-2. Descriptions of router types

Type	Feature	Description
DSL/Cable router	Up to 253 network users, Web site blocking, NAT routing, VPN pass-through, supports QOS, DHCP	This type of router enables two or more computers on a LAN to share an Internet connection. It can be used by students to access the Internet or used by a peer-to-peer computer lab for Internet access.
Router in small lab	Built-in security, device management software, WAN connection, 10/100 Mbps switch ports, VPN tunnels, support for WLAN, support for high-quality voice and video, integrated DSL	This type of router is good for a small lab connected to a university's network. It provides the integrated service that can run multiple services concurrently. It allows students to access online computer labs, provides voice service through VoIP, and supports secure WLAN access.
Router in medium lab	Built-in security, device management software, high-speed WAN, hundreds of VPN tunnels, wireless LAN services, support for high-quality voice and data integration, business-class DSL, Gigabit Ethernet ports, dial-up access services, multiple T1/E1/ADSL, Wi-Fi hotspot service, mobile radio Service, wireless infrastructure service	This type of router is good for a medium lab which consists of multiple networks. It supports a complete suite of transport protocols, device management tools, and advanced security and voice applications for wired and wireless deployments. It also supports various enhanced WLAN services.
Routers in large lab	Built-in security, device management software, built-in routed ports, built-in switch ports, DC power for IP phones, thousands of VPN tunnels, hundreds of VoIP phones, Gigabit Ethernet, full T3/E3 rates, high-speed WAN, wireless LAN services, Wi-Fi hotspot service, mobile radio service, wireless infrastructure service	This type of router is good for a large lab which consists of more than one autonomous system. It provides a highly secure and scalable platform for the enterprise-level network with concurrent T3/E3 connections. It supports a complete suite of transport protocols, device management tools, and advanced security and voice applications for wired and wireless deployments. It also supports various enhanced WLAN services.

Table 4-3. Descriptions of switch types

Type	Feature	Description
Workgroup switch	• Built-in security • Up to 16 10/100Base fast Ethernet ports • Router detection • Supports 10/100 Mbps LAN or 802.11g WLAN devices	This type of switch is designed for home network and a peer-to-peer computer lab for connecting several desktop and notebook computers and other equipment such as a printer or an external hard drive. It is also suitable to connect equipment in a group for small, medium, or large computer labs.
Wired switch	• Built-in security • Device management software • Up to 20 10/100/1000 Gigabit Ethernet ports or 48 10/100Base fast Ethernet ports • Router detection • Logically groups multiple switches • Automatically adjusts network speeds	This type of switch is an enterprise-level, stackable multilayer switch to provide high availability, security, and quality of service. It is used for connecting computers and other devices to a Gigabit Ethernet LAN or a fast Ethernet LAN.

continued on next page

Table 4-3. continued

Blade switch	• Built-in security • Device management software • Internal 1000BASE-T ports • External 10/100/1000 SFP ports • Supports copper and fiber (SX) SFP modules • External 10/100/1000 BASE-T ports • External console ports	This type of switch is designed to reduce cable complexity, save space, and improve security and management. It can decrease cable complexity up to 90%. It is needed for computer labs that adopt the blade server system.
Modular Ethernet switch	• Built-in security • Device management software • Layer 2 through layer 4 switching capabilities • Layer 2/3/4 services • 2-10GbE and 4GE uplinks • High-performance, high density access layer	This type of switch is designed for enterprise-level, small and medium-sized businesses, and metropolitan (metro) Ethernet customers by providing intelligent network services such as predictable performance, advanced security, comprehensive management, and integrated resiliency both in software and hardware. It is suitable for large computer labs.

Table 4-4. Descriptions of SAN types

Type	Feature	Description
Fibre Channel	• Supports hundreds of hard drives • Can be linked to hundreds of servers • Stores hundreds of TB data • Allows Gigabit transmission rate • Supports various operating systems such as Windows, Sun Solaris, HP-UX, IBM AIX, NetWare, and Linux • Provides disaster recovery services • Includes ports for Fibre Channel switches • Includes Fibre Channel host bus adapters	In a Fibre Channel SAN, servers and storage devices are connected by several Fibre Channel switches connected together to form a Gigabit-speed network. The network uses the SCSI protocol for communication between servers and devices. The servers in a Fibre Channel SAN network use special Fibre Channel host bus adapters (HBAs) and optical fibre. Fibre Channel SAN is faster than iSCSI. However, it is much more expensive.
Internet SCSI (iSCSI)	• Supports dozens of hard drives • Can be linked to dozens of servers • Stores dozens of TB data • Allows Gigabit transmission rate • Supports RAID structures • Supports various operating systems such as Windows, Sun Solaris, HP-UX, IBM AIX, NetWare, and Linux • Provides disaster recovery services • Backs up data over 100 MB/second • Includes Ethernet RJ45 connector ports	iSCSI is also called IP SAN which runs SCSI on a TCP/IP network. In iSCSI, the switches are Ethernet switches and the hosts use Ethernet network interface cards. IT professionals are familiar with IP network facilities, so it is much easier to deploy iSCSI. In general, iSCSI is slower than Fibre Channel SAN. For online computer labs, iSCSI is a good choice.

Table 4-5. Server functionalities
efficiently manage storage devices. SAN devices are commonly used for enterprise data storage. As Gigabit Ethernet is common now, more and more organizations accept the SAN technology. The use of SAN in computer labs is mainly for learning purposes, that is, to get students familiar with the e-commerce environment. SAN can also be used with learning management systems (LMS) and e-mail servers to

Computer Lab Model	Server Functionality
Peer-to-peer	None
Small lab model	User management
	Database management
	Network management
	Security management
	Patch and update management
	File sharing on local network
	Printing service
	Web systems management
	System backup and recovery
	FTP service for Internet file sharing
Medium lab model	User management
	Database management
	Network management
	Security management
	Patch and update management
	File sharing on local network
	Printing service
	Web systems management
	System backup and recovery
	FTP service for Internet file sharing
	Virtual private network service
	Groupware management for collaboration
	E-mail service
	Remote management service
	Application service management
	Network storage equipment management
	Job automation and scheduling
	Batch processing management
	Proxy service for safety, filter requests, and connection sharing

continued on next page

Table 4-5. continued

Large lab model	User management
	Database management
	Network management
	Security management
	Patch and update management
	File sharing on local network
	Printing service
	Web systems management
	System backup and recovery
	FTP service for Internet file sharing
	Virtual private network service
	Groupware management for collaboration
	E-mail service
	Remote management service
	Application service management
	Network storage equipment
	Job automation and scheduling
	Batch processing management
	Electronic Data Interchange
	Customer relationship management
	Audio/video service
	Proxy service for safety, filter requests, and connection sharing
	Clustering for high performance computing
	Support for distributed computing environment

store online course materials and backup e-mail messages. Table 4-4 lists the features of different types of SAN devices.

Building and running reliable online computer labs require careful design of network construction. The members of the development team need to have experience and adequate training in networking. For a large lab model, if the development team lacks skills or time to figure out technical details, you may consider outsourcing the tasks to a consulting company.

Selection of Software

The selection of software also depends on the requirements of teaching and hands-on practice. In Chapter II, when we discussed the Web-based teaching technologies, the functionalities of application software were described. In this chapter, we will focus on server-side software, which may include server operating systems and a variety of management tools. Before examining the software, let us give review some of the functionalities required by different types of lab models. For each computer lab model, the possible server functionality requirements are listed in Table 4-5.

The assignment of functionalities to each lab model is not strict. A server operating system is the software that can be used to accomplish most of these tasks. The selection of operating system software depends on the lab model and the functionalities to be supported by the server. Next, let us have a closer look at different types of server operating systems.

Server Operating Systems

The selection of server operating systems depends on the requirements of teaching and hands-on practice. A server operating system may not be the most popular one or may not have the best performance. However, if it is required by a technology-based course, then it should be installed in the lab. Often, multiple operating systems are installed in an online computer lab to support multiple technology-based courses. Table 4-6 describes the server operating systems that can be used in an online computer lab.

For many small colleges and universities, Windows and Linux are adequate to meet the needs of teaching and hands-on practice. With strong technical and financial support, a large university may have various operating systems installed in its computer labs.

Server Management Tools

While selecting server operating systems, you also need to decide if you need any of the utility tools to keep a server running smoothly or help the system administrator manage server operations. If so, you may want to find out if the server operating system software already includes the utility tools or you need to purchase some. Table 4-7 lists some commonly used utility tools supported by a server.

There are many other useful tools available for managing servers. The selection of these utility tools depends on the requirements, costs, and personal experience with these tools.

Table 4-6. Server operating systems

Server OS	Maker	Description
Windows	Microsoft	Windows Server 2003 is a widely used server operating system. It is relatively easy to learn due to the fact that it has a similar Desktop arrangement to Windows XP which is installed on most of the home computers. It is covered by the MSDNAA program. It has a very low price tag for an information systems department to install this type of server operating system in a computer lab. The technical support for this server operating system is also adequate. A lot of technical support teams are familiar with this type of operating system. Windows Server 2003 supports a large number of application software.
Linux	RedHat, Debian, Mandreke, Slackware, SuSE, and so on	Linux is an open source operating system. It can be used as a client-side and server-side operating system. It can be installed on computer systems sold by some major computer hardware companies such as Dell, IBM, HP, and Sun. As open source software, Linux is freely downloadable. However, some Linux distribution vendors sell Linux with a service package and supporting products which may cost hundreds of dollars. The Linux operating system can be used to support many technology-based courses. It is one of the popular operating systems installed in a computer lab. Technical support may vary from campus to campus. Often, on a small campus, Linux is less supported by the computer service department due to lack of knowledge about Linux. The application software is from different vendors and many of them are also open source software.
Solaris	Sun	Solaris is the operating system installed on SPARC and Ultra-SPARC workstations and servers. It can also be used on the x86 platform. Solaris performs better on high-end servers. For many major research universities, Solaris is used to support courses using UNIX operating systems, especially, graduate-level courses. Solaris is also popular in research labs.
AIX	IBM	AIX is a UNIX operating system developed by IBM for mainframes and workstations. It is easier to integrate AIX with Linux. The AIX operating system is often used in research labs and used to support UNIX related courses. There is a plenty of application software for AIX.
HP-UX	HP	HP-UX is a UNIX operating system developed by HP for Itanium servers and workstations. HP-UX is used by many universities, institutes of technology, academies, and schools. It is used to run large application software such as database servers, library systems, and e-mail.

Selection of Lab Access Technologies

Students should be able to remotely access an online computer lab from anyplace where there is an Internet connection. Based on the requirements of performance, as a lab designer, you can choose from several technologies for remote access to an online computer lab.

Virtual Private Network (VPN)

This solution creates a secure virtual private network between the client and the server. It can be designed to support a variety of computing platforms such as Windows, UNIX, Macintosh, PocketPC, and Linux. It provides protection for the Web, e-mail, databases, and file transformation. The VPN technology may include the following components.

Table 4-7. Server management utilities

Utility	Description
Remote desktop	It is a utility that allows a user to access the desktop of a remote computer from his/her own computer.
Terminal services	A utility that can be configured to allow a user from a remote computer to access applications or data stored on the current server.
FTP service	It is a service used to transfer files from one server to another across the Internet.
Server Side Includes (SSI)	SSI allows a user to add content dynamically to a Web page without having to recreate the entire page.
Proxy SSL encryption	SSL stands for Secure Sockets Layer. This service provides data encryption and authentication of servers or clients.
Firewall protection	A firewall blocks hackers and other unknown threats that can cause devastating damage to your server.
Server re-image	Server re-image allows a user to reset the server to default settings.
Secure Shell (SSH) access	Secure Shell (SSH) access provides strong authentication and secure communications.
GeoTrust QuickSSL Premium Dedicated SSL Certificate	It provides safety for information exchange over the Internet and shows visitors that their transactions are secure.
Deployment, migration, and management tools	Tools that are used to deploy, migrate, and manage server operating systems and applications.
System monitoring	A system monitoring utility allows users to observe both overall system performance and single-performance of an execution.
Cluster management	This utility can be used to automate regular cluster management tasks to reduce administration workload.
Event automation	This utility can use server events to automatically generate e-mail.

continued on next page

Table 4-7. continued

Batch processing	It is used to conduct a group of computing tasks at one time.
Data center management	This utility provides programming and operational support for managing data storage, performance tuning, and disaster recovery.
Disaster recovery	It can be used to recover from a natural disaster or malicious intent. Disaster recovery includes system replication and backup/restoration.
Job scheduling	A job scheduling utility can be used to assign computing tasks to a set of resources.
Performance management	This utility measures, predicts, and optimizes system performance over time by monitoring performance indicators such as response time, resource utilization, demand and contention, and queue length.
License management	This utility enforces software license agreements.
Network management	This utility is used to remotely or locally monitor and configure networks.
Patch and update management	It checks firewalls, automatic updates, and zone scanning for missing updates for the system and application software.
Printing management	This utility is used to control and audit printer activities. It also monitors printing tasks on computers.
Storage management	This utility can be used to monitor and track the performance of network storage equipment such as SAN. It can also be used to manage capacity utilization and availability of file systems and databases.
User and computer management	This is a utility that maintains a user's or a computer's preferences and privileges. Often, the management tasks can be done through a directory service.
Web systems management	Utilities of Web systems management are used to deploy scalable e-commerce applications and manage Web data centers and Web content.

- Client software which is installed on each student's computer for remote access.
- A VPN server or network access server (NAS) for dial-up service or remote access service.
- VPN security and management software.

There are two types of VPN, remote access VPN and site-to-site VPN.

- **Remote access VPN:** It is also called a virtual private dial-up network (VPDN). It connects client/student computers from remote locations to the LAN in an online computer lab. In the lab, a server should be set up as the remote access VPN server. On the client side, VPN client software should be configured to communicate with the VPN server. The VPN client software may be already included in some operating systems such as Windows XP. VPNs provide secure, encrypted connections between a student and the private network in a computer lab through the Internet.

- **Site-to-site VPN:** This solution is used to provide students with remote access to network resources through a university's intranet. The intranet may be more secure but it is limited to the university's network. It blocks students' access to other resources on the Internet.

The cost of VPN server software is small. Some of the server operating systems already include a VPN, such as Windows Server2003. A VPN can also be implemented by an equipment called a concentrator which is built for a remote access VPN. A VPN may also be included in a router. A VPN is a good choice for computer lab models from small to large. As mentioned before, a server for a large computer lab model can handle up to thousands of VPN tunnels.

Citrix

Citrix is a suite of tools designed to allow multiple students to run multiple applications on the Citrix server at the same time. A Citrix server sends screen shots to client computers and, in return, the client computers send keyboard input and mouse movements back to the Citrix server. The following are some functionalities that are valuable to online computer labs.

- Citrix can be used to enhance interactivities between students and instructors through the use of real-time chat, virtual whiteboard, screen sharing, and Web page push.

- Citrix can push a list of application software to students' client computers for local processing. You can also manage the application software installed on a student computer without visiting the student's home.

- Citrix supports heterogeneous computing environments. It allows various operating systems such as Windows, Mac, and Linux running on different computing platforms to access a Citrix server. Students can access a Citrix server through a dialup phone line, T1 line, ISDN line, Wireless, and cell phone.

- Multiple servers can be installed on a single Citrix server. This will take care of multiple courses that require different operating systems.

Although there is some cost for Citrix, it is an efficient solution for a small or medium-sized computer lab.

Terminal Services

Terminal Services is a remote access tool developed for Windows operating systems. This technology is originally created by Citrix. Students can access a server through the Remote Desktop Connection utility provided by Windows XP or Terminal Service Client software. The Terminal Services tool is included in Windows Server 2003. Besides the money saving aspect, Terminal Services allows students to access the server from anywhere. If a student has a non-Windows operating system on his/her client computer, he/she needs to obtain simple interface software to run Terminal Services. If a lab has a Windows-based server, students can remotely access the server through Terminal Services and do not have to log on as administrators. Terminal Services is a low cost solution for a small lab model. However, the performance of Terminal Services is not as efficient as Citrix.

Remote Access through Application Software

A lot of application software such as database and e-mail allows a user to remotely access them without a user account in the operating system. For example, students can log on to an Oracle database server through the Internet and perform hands-on practice if they have an account for the Oracle database. Microsoft SQL Server can do the same. If a course only needs a database to support its hands-on practice, there is no need to use other remote access methods.

Remote Access through Virtual Machine

Another remote access method is through a virtual machine which is a software package used for software testing. A virtual machine simulates the function of hardware with software. Once a virtual machine is installed on a server, students can access the virtual machine through the Internet. After logging on to the virtual machine, the students can perform their hands-on practice on the virtual machine instead of on the physical server. One can create multiple virtual machines, each with a different operating system. This will allow multiple operating systems to run simultaneously. It is especially useful for courses that need the server side hands-on practice. Since a virtual machine is isolated from the physical server, it provides a secure computing environment. The disadvantage of a virtual machine is its performance. Especially when multiple virtual machines are running on the same physical server, the performance of these virtual machines is significantly decreased. Since each virtual machine needs its own memory and storage space, it is required that the physical server have a large amount of RAM and a very big hard drive. Based on its performance, a virtual machine is a good choice for a small computer lab model.

Conclusion

In this chapter, we have discussed some strategies for developing an online computer lab. We first examined various computer lab models. Detailed descriptions and figures were used to illustrate each of the online computer lab models. This chapter has looked into the details in solution design. We have investigated hands-on practice activities and translated them into a series of technical terms. A logical design perspective of each type of online computer lab was given to illustrate the design process.

Along with the discussion, some recommendations were given for each of the computer lab models. As a task of physical design, to help specify servers, various server platforms were discussed.

This chapter has discussed the selection of some major network equipment such as routers and switches. It examined the SAN technology. For the selection of software, this chapter has focused on server operating systems. To help with the selection of server operating systems, the chapter first investigated the tasks to be done by a server in a computer lab. Then, it provided information about some commonly used server operating systems and the roles they can possible play in a computer lab. The discussion of some commonly used utility tools was also given. This chapter has listed the features of some commonly used remote access software which allows students to remotely access the online computer lab. Suggestions on the possible use of some remote access software were also given in this chapter.

Once the tasks covered in this chapter are accomplished, the blue print of the online computer lab is created and the lab is ready to be physically built. In the next chapter, we will start to consider issues on developing and implementing the online computer lab.

References

Andersen, S., McDermott, D., Dial, J., & Cummings, C. (2005). Infrastructure architectural capabilities. Retrieved April 5, 2007, from http://msdn.microsoft.com/library/default.asp?url=/library/en-us/dnbda/html/InArchCap.asp

Bollapragada, V., Khalid, M., & Wainner, S. (2005). *IPSec VPN design*. Indianapolis, IN: Cisco Press.

Chevance, R. J. (2005). *Server architectures: Multiprocessors, clusters, parallel systems, Web servers, storage solutions*. Burlington, MA: Digital Press.

Chong, F., Clark, J., Morris, M., & Welsh, D. (2005). Web service health modeling, instrumentation, and monitoring: Developing and using a Web services health model for the northern electronics scenario. Retrieved April 5, 2007, from http://msdn.microsoft.com/architecture/learnmore/learnimpdes/default.aspx?pull=/library/en-us/dnbda/html/MSArcSeriesMCS6.asp

ePlanetLabs (2005). MCA Microsoft Certified Architect study guide. Retrieved April 5, 2007, from http://eplanetlabs.com/mca.html

Erl, T. (2005). *Service-oriented architecture: Concepts, technology, and design*. Indianapolis, IN: Prentice Hall.

Groth, D., & Skandier, T. (2005). *Network + Study Guide* (4th ed.). Alameda, CA: Sybex.

Guo, X., Katz, L. & Maitland, M. (2002). An evaluation of an online anatomy course by laboratory instructors: Building on instructional design. In G. Richards (Ed.), *Proceedings of World Conference on E-Learning in Corporate, Government, Healthcare, and Higher Education 2002* (pp. 1552-1554). Chesapeake, VA: AACE.

Lewis, J., Wimberg, J., Kinuthia, W., Guest, J., & Jackson, K. (2003). Distance learning faculty development workshop using the ADDIE model. In C. Crawford et al. (Eds.), *Proceedings of Society for Information Technology and Teacher Education International Conference 2003* (pp. 382-383). Chesapeake, VA: AACE.

Marcraft International (2005). *Server+ Certification Exam Cram 2 (Exam SKO-002)*. Indianapolis, IN: Que.

Meliksetian, D.S., et al. (2004) Design and implementation of an enterprise grid. *IBM Systems Journal*. Retrieved June 7, 2006, from http://www.findarticles.com/p/articles/mi_m0ISJ/is_4_43/ai_n9544042

Microsoft. (2006). MSDN Solution Architecture Center. Retrieved April 5, 2007, from http://msdn.microsoft.com/architecture

Ransom, M., et al. (2005). The solution designer's guide to IBM on demand business solutions. IBM Redbooks. Retrieved June 5, 2006, from http://www.redbooks.ibm.com/abstracts/sg246248.html?Open

Recore, M., Laurenson, J., & Herrmann, S. (2003). *Cisco Internet architecture essentials self-study guide: Cisco Internet solutions specialist: Infrastructure for eBusiness services.* Indianapolis, IN: Cisco Press.

Stearns, B., Zhu, X. H., Cui, M. Z., Shu, B., Xu, Y. Z., Li, X., Li, M., & Qu, F. (2004). E-business globalization solution design guide: Getting started. IBM Redbooks. Retrieved June 5, 2006, from http://www.redbooks.ibm.com/abstracts/sg246851. html?Open

Wiehler, G. (2004). *Mobility, security and Web services: Technologies and service-oriented architectures for a new era of IT solutions.* Hoboken, NJ: Wiley-VCH.

Section III

Development of Online Computer Labs

Chapter V

Server Development for Online Computer Labs

Introduction

The previous chapters have covered the important design stages in developing an online computer lab. Chapter III discussed the issues on understanding the requirements and planning the development of online computer labs. In Chapter III, we discussed the issues about collecting information to get the perceptions of the future online computer lab from the users' points of views. In Chapter IV, the users' perceptions were translated into technical terms. The lab activities were represented by a flowchart. Computer lab models representing different types of labs were examined. Physical design related topics such as strategies for selecting technologies were also discussed in Chapter IV.

In the next few chapters, we will discuss issues related to the implementation of online computer labs. Once the technologies are specified for the components in a computer lab model, the next task is to put these technologies together to construct a computer lab physically to support teaching and hands-on practice. In this chapter, the issues about server implementation will be addressed.

Based on the activities required for teaching and hands-on practice, the servers should be configured to support these activities. The developers need to make a decision on the configuration of each building block in the server architecture. To accomplish this task, the developers need to specify the speed of the central processing units (CPUs), the size of the hard drives, and the size of the memory for each server. They will also need to specify how the servers communicate with each other and with client computers in the network, how to deploy the servers, and how the peripheral equipments can be attached to the servers.

This chapter will first address the server-side hardware configuration issues. After the discussion of the hardware configuration, we will deal with the server-side software configuration issues. We will discuss some general procedures of configuring the hardware and software with the lab design objectives in mind. Lastly, we will discuss some configuration issues about various special purpose servers.

Background

Properly configuring a server to support the network-based computing environment is a critical step in an information system. Servers are widely deployed in various organizations from small companies to global enterprises. For online computer labs, except the peer-to-peer lab model, the other lab models all need the support of various servers. For server development, Boer (2002) discusses a delivery server infrastructure developed for distributing sophisticated multimedia content. This delivery server infrastructure provides a platform independent framework to support interactive distance learning. Ayala and Paredes (2003) report the development of a learner model server that supports a personalized learning environment. This kind of server can provide different sets of applications based on learners' interests.

There are very few instructional design studies on the development of a Web server which is one of the special purpose servers mentioned in this chapter. In their paper, Fansler and Riegle (2004) discuss a data-driven online instructional design analytics system. In their project, a Web server is involved in the design process. As described in the above, the studies about servers focus on delivering the course content to support instruction. In this chapter, not only shall we cover the servers used to deliver the course content, but also cover the servers that manage the networks in an online computer lab and the servers that support remote access.

As mentioned in the previous chapter, there are various server platforms such as x86, IBM POWER, Sun SPARC, and Itanium 2. Servers with the x86 platform are popular for the small or medium computer lab model. Part of the reason is that technical support teams and users are more familiar with the x86 platform. It takes less effort for a technical support team to implement the computer lab and it requires less training for the faculty members and students to use the lab. The x86 platform supports most of the operating systems. The configuration of the x86 platform often includes the specification of hardware components such as motherboard, memory, or hard drives (Buchanan & Wilson, 2000). A large computer lab model may use any of the platforms to implement the computer lab. For a high-end computing environment, POWER, SPARC, or Itanium 2 should be considered. Although the x86 platform is sufficient for a small or medium online computer lab, many online labs still have the high-end computing platforms for learning purposes since those platforms are commonly used by larger companies. The POWER platform is developed by IBM. It supports operating systems such as IBM AIX and MacOS. It supports the 64-bit computing environment, which is a desired feature for the virtualization of processors, storage, and networks (Vetter et al., 2005). Another 64-bit platform is the SUN SPARC platform (SPARC International, 1994), which supports operating systems such as Linux and Solaris. To run both the Windows and UNIX operating systems, Intel and HP jointly developed Itanium 2 (Intel Corporation, 2006). Itanium 2 supports Windows, Linux, and UNIX. Beyond the specific platforms mentioned above, in-depth knowledge about various computing platforms is often rooted in computer architecture related topics (Englander, 2003).

In the previous chapter, we have looked at several commonly used server operating systems such as Windows Server and Linux. Among the server operating systems, Windows and Linux are more popular for the small and medium computer lab models. In this chapter we will discuss some configuration issues about these server operation systems. The configuration of Windows Server 2003 server is relatively easy to handle (Minasi, Anderson, Beverridge, Callahan, & Justice, 2003). Configuring the Linux operating system may take longer time for a technician who is new to the operating system, but many experienced professionals like Linux since they can have more freedom to configure the system the way they want (Hunt, 2002). The UNIX based systems vary from one vendor to another; customers often need to work with a vendor's consulting team to figure out the system configuration and cost. The configuration of AIX can be accomplished by following the guidelines provided by Vetter, Pruett, and Strictland (2004). Lab developers can also configure the HP-UX operating system by following the guidance provided by Poniatowski (2005). A large computer lab may use multiple server operating systems. Sometimes, a large university may have a joint program with some of the major hardware or software companies. In such a case, these companies will provide strong technical support and training for their products.

About the server-side hardware, servers can be built as stand-alone servers; that is, each server is a minitower box and is connected to other servers through a network. To simplify server development and management, servers can also be plugged in a rack frame (Lowe, 2001) or a blade enclosure (Chapa & Radu, 2006).

Next, we will further look into the issues on the implementation of server hardware and software.

Server Hardware Configuration

Chapter IV discussed the issues of selecting suitable servers for a computer lab. After we have done the investigation and purchased the servers, the next step is to install and configure the servers according the requirements of teaching and hands-on practice. First, let us discuss the issues related to the requirements of servers for each online computer lab model.

Small Computer Lab Model

A small online computer lab needs the servers to provide services in lab management, remote access, application support, Web, and security. For a small computer lab, you may need to install the following servers.

* Several servers may be used to support students' practice of networking and system administration. Assume that students in a class are divided into several groups and each group has its own server. On these servers, Linux or Windows operating systems are installed. Students are given the administrator's privilege. Therefore, these servers are in an unstable condition; that is, they may be reconfigured frequently based on the teaching and hands-on practice requirements.

* Servers are needed to support application software such as application development packages like JDK and Visual Studio .NET. The lab may also need servers to host special purpose servers such as application servers, e-mail servers, and database servers to support database application development. The servers and the application servers hosted by these servers should be kept in a stable condition.

* Servers are needed to support database administration. Each server will be used to support a group's activities. It is required that these servers themselves are in a stable condition but the database servers may be in an unstable condition because the database servers are used to support ongoing hands-on practice of database administration by students. Therefore, these database servers should not be used to support courses like database application development.

* Servers are needed to host Web servers. If a Web server is used to support Web application development, it should be kept in a stable condition. On the other hand, if a Web server is used for hands-on practice of Web service development and administration, they may be in an unstable condition.

* A server is needed to support the VPN server for remote access to the online computer lab.

Based on these needs, we will have at least three types of servers. The first type includes servers that are in an unstable condition; the second type includes both stable servers and stable special purpose servers hosted by those stable servers; the third type includes stable servers and unstable special purpose servers hosted by these stable servers. For example, if students are divided into four groups, then we will need four unstable servers, one stable server to support application development, one or more stable servers to support unstable special purpose servers, and one stable server to host the VPN server. As an example of another situation, in some network administration courses, a class of 12 students may need 12 or even 36 unstable servers; these servers can be a set of standard-alone servers, or servers hosted by a rack system or by a blade system.

It is important that the stable server supporting application development be powerful enough to support the server operating system and multiple applications which will share the CPU, memory, and hard drive on the same server. For an unstable server, its job is to host an operating system. Therefore, you may use a less powerful computer for this kind of server.

To specify each type of server mentioned above, let us consider some examples to demonstrate the configurations of servers. The examples below are Intel or AMD based systems, which belong to the x86 platform. For the purpose of online computer labs, entry- or midrange-level servers should be adequate for most of the cases. The following are examples of typical structure configurations for stand-alone, rack systems, and blade systems.

Stand-Alone Server

If you decide to use standard-alone servers to construct an online computer lab, you can consider the following configurations. For an unstable server, it will depend on the type of operating system to be installed on the server. For Windows Server 2003 or Red Hat Enterprise Linux ES, the following configuration of an unstable server should be adequate.

- CPU: One Pentium III processor at 550 MHz.
- RAM: 512MB single ranked DIMM.
- Hard drive: One 20GB IDE hard disk.
- CD/DVD: 24X CDRW.
- A floppy drive.
- Graphic and network cards.

The above requirements are for a low-end server. You can even find some surplus computers that will match the above requirements. If you want to install multiple operating systems on one computer or to have better performance, you can consider using the following configuration.

- CPU: One Intel IV processor at 2.8 Ghz/1MB cache.
- RAM: 1GB single ranked DIMM.
- Hard drive: Two 80GB, Serial ATA, 7.2K RPM hard disks.
- CD/DVD: 48X, CDRW/DVD.
- A floppy drive.
- Graphic and network cards.

The above configuration is also good for a stable server to host an unstable database server or e-mail server.

Due to the fact that some of the servers are in an unstable condition, you may consider using the virtual machine option. One or more virtual machines can be installed on a stable server. Using virtual machines can drastically reduce maintenance of these unstable servers. Before each semester starts, you do not need to reinstall the server operating system on each computer. All you need to do is to paste a copy of the virtual machine file to each computer. The disadvantage of using virtual machines is that virtual machines have slower performance compared with the host server. Also, a virtual machine requires a lot more RAM. The following configuration can be used to host seven virtual machines.

- CPU: Two Intel 3.2GHz/2MB Cache Xeon processors.
- RAM: 4GB dual ranked DIMMs.
- Hard drive: Two 80GB Serial ATA, 7.2K RPM hard disks.
- CD/DVD: 48X, CDRW/DVD.
- A floppy drive.
- Graphic and network cards.

The above configuration is also good for a stable server to host several stable special purpose servers. As mentioned in the previous chapter, a high performance stand-alone server may support up to 64GB RAM and 4TB hard disks. Thus, it is possible to construct an entire computer lab on a single server.

Rack System

As described above, a small computer lab model may have several servers. To efficiently manage the servers, the servers can be made smaller in size and be plugged in a rack frame. Each server on a rack system has similar functionalities to a standalone server. For the specifications of a rack system, let us consider the following example which demonstrates a typical configuration of a low-end rack system.

In this example, an entry-level rack system is configured. We assume that six entry-level servers are plugged in a 22U rack frame. Each server has the following configuration.

- CPU: Two Intel Xeon 3.2GHz processors.
- RAM: 1GB dual ranked DIMM.
- Hard drive: One 80GB Serial ATA, 7.2K RPM hard disks.
- CD/DVD: 24X CD-ROM.
- Cache: 1MB Level 2 cache.

This entry-level rack system is powerful enough to support most of the teaching and hands-on activities in a small computer lab. It can support multiple special purpose servers and run several virtual machines simultaneously. A 22U rack frame plus the installation kit and services cost about $3,000. The total cost for an entry-level rack system is about $20,000. When you make a budget plan for a small computer lab, a rack system like this one can be considered if there is enough funding for the online computer lab project.

Blade System

Servers can be made even smaller and be plugged in a blade enclosure. A blade system takes much less space and is much easier to manage. The following is a configuration example for an entry-level blade system. Suppose that six blade servers are plugged in a blade enclosure. Each of the blade servers has the following configuration.

- CPU: Dual Core AMD Opteron 1.8GHz processor.
- RAM: 1GB SDRAM.
- Hard drive: One 72GB Serial ATA, 7.2K RPM hard disk.

An 8U blade enclosure with power supply units costs about $5,000. The total cost for this entry-level blade system is about $20,000.

Both the rack system and blade system can drastically reduce the computer lab space. An entire system can be kept in an office. If a computer lab is built for an online class only, you should give the rack system or blade system serious consideration. However, if a computer lab is shared with face-to-face classes, you need to consider using stand-alone servers since students in a face-to-face class need space for their lab work and each of them needs a monitor, keyboard, and mouse for hands-on practice.

Medium Computer Lab Model

In a medium computer lab, servers are used to support group activities. They are also used to construct client-server architectures and support remote access. In addition to the servers used in a small computer lab, a medium computer lab may require the following additional servers.

- Several servers may be needed to support networking and system administration. For a medium computer lab, these servers should be able to handle directory services such as Microsoft Windows Active Directory or Lightweight Directory Access Protocol (LDAP).

- Servers are used to support client-server computing. Each client-server structure includes a server as the backend and a midtier server which serves as an application server.

- Servers may be used to support hands-on practice for server administration. Each group of students will work on a domain that contains at least two servers.

- A server can be configured as a proxy server that can be used as a gateway which separates the computer lab from the Internet. It can also be configured as a firewall server to protect the computer lab.

Again, servers can be grouped as unstable servers, stable servers, and unstable special purpose servers installed on a stable server. In addition to the servers mentioned for the small computer lab model, among the above additional servers, a directory server can be used as a stable server that an instructor can use to manage computers and users. The servers used in a client-server system could be unstable servers. The servers used to construct domains are also unstable servers on which students can play the role of an administrator. The proxy server has to be a stable server. Again, these servers can be standard-alone servers, a rack system or a blade system. These servers can be x86 based servers, or other platform based servers. In the following, some configuration examples are given for different types of servers.

Stand-Alone Server

The configurations of x86 based stand-alone servers are similar to those mentioned for the small computer lab model. Sun also produces stand-alone servers. However, it is often that Sun UNIX based servers are built for a rack system or for a mainframe.

Rack System

A rack system can be an x86 based system or UNIX based system. For the UNIX based servers, we have IBM POWER servers, Sun SPARC server, and HP Itanium 2 server. In the following, let us take a brief look at the configurations of these servers on different platforms.

For an x86 based rack system, you may find that the following configurations will be adequate to meet most of the needs of a medium computer lab. Assume that 21 servers are plugged in a 42U rack frame. One of the servers is used as the management server and the other 20 servers are used to support group activities, practice of system administration, and client-server computing. The configuration for each server is specified below.

- CPU: Two dual-core AMD Opteron processors.
- RAM: 4GB dual ranked DIMM.
- Hard drive: Two 73GB Serial ATA, 7.2K RPM hard disks.
- CD/DVD: 24X CD-ROM.

The cost of such a rack system may be well over the budget limit at a small college or university. In such a case, it is necessary to think about an alternative budget plan, such as using some of the older stand-alone servers that work side by side with a rack system.

If a technology-based course requires an IBM POWER server, the following is a possible configuration that will meet the requirements of the server in a medium computer lab.

- CPU: Two POWER5 1.5GHz processors.
- RAM: 16 GB RAM.
- Hard drive: Four 74GB 15K hard disks.
- Network card: Two 10/100/1000Ethernet ports.

Sun also produces entry or midrange UNIX based servers. One of them is the general purpose Sun Fire V490 server. The following is the configuration that can meet most of the requirements of the server for a medium lab.

- CPU: Two 1.5GHz UltraSPARC IV+ processors.
- RAM: 8 GB RAM.
- Hard drive: Two 246GB 10000 RPM FC-AL hard disks.
- Network card: Two 10/100/1000Ethernet ports.

HP makes Itanium 2 servers. If an Itanium 2 server is required by a technology based course for teaching and hands-on practice, the following configuration is common for an entry-level or midlevel Itanium 2 server.

- CPU: Two Intel Itanium 2 processors.
- RAM: 8GB PC2100 ECC registered DDR266A SDRAM.
- Hard drive: Four 36GB 15K RPM hard disks.
- Network card: Two 10/100/1000Base-TX LAN ports.

These UNIX based servers can either run server operating systems developed by individual companies or a version of the Linux operating system.

Blade System

A blade system can also be built to support a medium computer lab. A blade system can be an x86 or UNIX based system.

For an x86 based blade system, the following is a possible configuration for a medium computer lab. Suppose that there are 20 blades and each of the blades may have:

- CPU: Two dual-core AMD Opteron 2.2GHz processors.
- RAM: 2GB SDRAM.
- Hard drive: Two 72GB Serial ATA, 7.2K RPM hard drives.
- Embedded Gigabit Ethernet.

For a medium computer lab, you can also use a UNIX based blade system if it is required. The configuration of a UNIX based blade system is about the same as that of an x86 based blade system except that different vendors use different CPUs in their systems. UNIX based blade systems from different vendors also run their own versions of the UNIX operating system or Linux operating system.

In addition to producing the x86 based blade system, IBM also offers blade servers based on their own POWER processors. The following is a configuration of a blade that may satisfy the requirements of the server for a medium computer lab.

- CPU: Two 64-bit IBM PowerPC 970MP up to 2.70GHz processors.
- RAM: 8GB of main memory.
- Hard drive: Two 72GB of internal hard disks.
- Embedded Gigabit Ethernet.

A Sun blade system allows SPARC and x86 based servers to be plugged in the same blade system for flexibility. In a Sun blade system, each blade can be configured with a similar size of RAM, hard drives, and network ports to the above configuration, except that the Sun blade system uses the UltraSPARC IV, UltraSPARC III, or UltraSPARC IIIi CPUs. Similarly, HP offers Itanium 2 blade systems, which use 64bit Itanium 2 processors. Some of the Itanium 2 blade systems can support up to 128 CPUs working simultaneously.

Redundant Array of Inexpensive Disks (RAID)

Servers in a medium computer lab are often installed with multiple hard drives. To improve reliability and performance, multiple hard drives can be configured as a RAID system (Patterson, Gibson, & Katz, 1989). Courses such as system administration and database management may cover the content related to RAID. If you configure the hard drives as a RAID system, a server can be used to support teaching and hands-on practice required by these courses. The commonly used RAID levels are RAID 0 (striping), RAID 1 (mirroring), RAID 5 (striping with parity), and RAID 10 (striping and mirroring). In the following, let us see how these RAID levels work.

- **RAID 0:** The RAID 0 system improves performance through disk striping, which is a process that divides a set of data into blocks called stripes. The stripes are then distributed to different hard disks to prevent I/O bottlenecks. Although RAID 0 can improve the system's performance, it does not have hardware fault tolerance. If one of the disks fails, the entire logical disk will fail with it.
- **RAID 1:** The RAID 1 structure provides hardware fault tolerance by duplicating data to multiple disks; that is, there are multiple copies of the same data

stored on different hard disks. In such a way, even when a disk fails, its duplicate (mirror) can continue to work, so the data will not be lost. The wonderful fault tolerance provided by the RAID 1 structure comes with a cost; that is, RAID 1 decreases performance and it costs more. For example, if two copies of the same data are saved on two different hard disks, each operation has to be carried out twice and the cost on the hard disks doubles.

- **RAID 5:** The RAID 5 system has good reading performance and fault tolerance by way of using parity information. It provides fault tolerance without costing twice as much of the disk space. If any of the disk drives fails, parity can be used to recreate the lost data and store them on another disk drive. It only uses a fraction of a disk drive to store parity information. RAID 5's reading performance is good because it uses disk striping. On the other hand, its written performance is slightly degraded. The RAID 5 system is a cost-efficient solution for a mission-critical system.

- **RAID 10:** The RAID 10 structure is a combination of RAID 0 and RAID 1. It has good performance like RAID 0 and provides fault tolerance like RAID 1. However, also like RAID 1, the cost of RAID 10 is higher. To sum up, when you use RAID 10, data are striped to multiple hard drives and each stripe is mirrored in other hard drives.

Based on the hard drive structure, a RAID system can be implemented with technologies such as fibre channel, SCSI hard drives, and Serial ATA or IDE hard drives. The following are the brief review about the RAID systems constructed with these technologies.

- **Fibre channel RAID:** A fibre channel RAID is often used with Storage Area Network (SAN) equipment. It is compatible with SAN technologies such as switch fabric. A fibre channel RAID is ready to be added to an existing SAN through a fibre channel host bus adapter (HBA) or a fabric switch. A fibre channel RAID system is configured to offer high performance and provide no-single-point-of-failure data protection.

- **SCSI RAID:** SCSI RAID is the most commonly used RAID system. The SCSI structure is often built in a server computer. A SCSI RAID system is less expensive than a fibre channel RAID system. Its performance and reliability are not too far behind those that a fibre channel RAID system can offer.

- **Serial ATA or IDE RAID:** A Serial ATA or IDE RAID system is much less expensive and has a relatively large capacity. Serial ATA or IDE hard drives are usually installed in PCs. These types of hard drives are widely available to implement RAID systems. When compared with other RAID systems, Serial ATA or IDE RAID has a lower price per megabyte ratio. The disadvantage of

Serial ATA or IDE RAID is its performance. Both the fibre channel RAID and SCSI RAID have superior performance and reliability when compared with Serial ATA or IDE RAID.

In a computer lab, there are two ways to use a RAID system. Students can create, configure, and manage a RAID system. In such a way, a RAID system is an unstable system and should not be used to support application software. The other way to use a RAID system is to use a RAID system to support special purpose servers. In this case, the RAID system must be reliable and have good performance. In fact, a RAID system can be implemented in a small, medium, or a large computer lab whenever a course requires a RAID system for teaching and hands-on practice.

Large Computer Lab Model

A large computer lab may be a combination of multiple medium computer labs distributed over multiple campuses. It may have a very complicated network structure and high-end servers, or it is simply a group of medium computer labs connected through the Internet.

The configuration of a large number of servers to make them to work together can be a strenuous task. The following are some general guidelines to accomplish the task.

- Specifying how many servers will be installed and where they will be installed.
- Specifying the roles to be played by these servers. For example, some of the servers will be domain controllers and some of the servers will be used to host application software. You may also need to specify the relationships among these servers.
- Configuring the servers based on their roles in the large computer lab.
- Testing the servers to make sure they actually work and can communicate with one another from remote locations for centralized control and remote management.

Due to the fact that a large computer lab can be distributed over multiple locations, one can implement a cluster or grid computing environment to support the distributed computing related courses. Both cluster computing and grid computing consist of networked computers that work together like a single computer. Both of them are constructed on the multitier client-server architecture. The middle tier is used to implement computation logic and manage messages for parallel computing.

Table 5-1. Differences between cluster and grid

Cluster Computing	Grid Computing
Computers are physically networked in a local area network.	Computers are physically networked in a wide area network or the Internet.
Computers work together to handle computationally intensive tasks.	Grid computing is used to handle distributed computing load and resource sharing.
Computers in a cluster are in the same platform and have the same operating system.	Computers in a grid can be in different platforms and have different operating systems.
A cluster computing environment is relatively stable.	A grid computing environment is more dynamic. The number of computers and resources may vary from one task to another.

However, there are some differences between cluster computing and grid computing. Table 5-1 shows the differences between the two.

There could be over 1,000 computers participating in a cluster. In such a case, the use of blade systems is a better solution. Multiple blade enclosures can be plugged in a rack frame or blade servers can be plugged directly in a specially designed cabinet. It is much easier to implement a cluster in a blade cabinet or a rack frame. It requires less network cabling and has better performance. Similarly, a grid can be constructed on multiple blade systems or rack systems distributed over multiple campuses. In the following, we will take a quick look at some high-end blade or rack systems that can be used to construct cluster or grid computing environments.

Cluster Computing

Clusters are used to solve computationally intensive problems in business and research institutions. To construct a cluster computing environment, you may need hundreds of servers and cluster management software. A high-end blade system constructed for a cluster may be configured as below.

- Cluster management software may be sold with the blade system. The management software can be Linux based, Windows based, or software developed by a vendor to match the operating systems on the blade servers.

- Each blade server can have four 64-bit processors from Intel, AMD or processors from IBM, Sun, or HP.

- Each blade can handle as much as 64GB of memory.

- A blade system can be accompanied by a storage server that can handle as much as 48TB of hard disks.

- A rack frame or blade cabinet are constructed with high performance optical or Gigabit Ethernet cluster interconnection.

Even with hundreds of server blades and dozens of storage blades, a blade system takes only a tiny area to run the cluster. Clusters can have high computing performance that no stand-alone computers can offer. Clusters are also more cost-effective than single computers of comparable speed or reliability.

Grid Computing

Unlike a cluster which is usually built in a cabinet or a rack frame, a grid is distributed across multiple locations. A number of companies, organizations, and universities have employed grid computing to support e-commerce (Joseph & Fellenstein, 2003). The key component of grid computing is the grid management software used to distribute the computing load and assemble the results of computing from hundreds or thousands of computers. The grid management software may be sold separately or sold with the hardware. There are a number of hardware vendors that provide the grid management software while selling the servers. Sun and IBM are such companies that offer both hardware and software for grid computing.

Uninterruptible Power Supply Configuration

During the operations of computer labs, a power outage can happen. To protect the servers, uninterruptible power supplies (UPSs) are used. There are two types of UPSs, the standby UPS and the generator-based online UPS. The following are some descriptions for each type of UPS system.

Standby UPS

This type of UPS is often a battery supported UPS. When the primary power voltage drops to a certain level, the UPS automatically switches the power to a backup battery. When the primary power is restored, the UPS switches back. The switchover time usually takes less than four milliseconds. Most of the UPS systems sold on the market are battery based systems. The advantage of a battery based UPS is that they are less expensive and can be used for various electronic devices of different sizes. The disadvantages are that its lifespan is short and the cost for maintenance and disposal is high.

Generator-Based Online UPS

For a large and mission critical server structure which requires several days of holding time, a battery based standby UPS is not adequate. The protection of this type of system needs help from a generator. The whole server structure is powered by a battery. During the operation, the battery is charged by the alternating current (AC) power if the voltage is below a certain level. During power failure, the voltage of the battery will eventually go down to a certain level. Then the standby generator starts the charging process until it runs out of diesel fuel. Thus, for a mission critical computer lab where no downtime is allowed, the generator-based UPS is a solution.

Choosing a suitable UPS system is important for running a successful online computer lab. The choice of a UPS for your equipment depends on the power consumption of the devices and the holding time during a power outage. To help users decide which UPS to purchase, some of the UPS vendors provide an interactive tool to help them to choose the right UPS. For example, the APC (2006) Web site provides a calculator for selecting UPS. From the Web site, a UPS buyer will be asked to enter equipment information, then the power consumption will be calculated and the suggestion of a proper UPS system will be given.

Server Software Configuration

In this paragraph, we are going to discuss issues related to the configuration of server operating systems. The configuration of software is also based on the requirements of teaching and hands-on practice. In the previous paragraph, the requirements for each type of computer lab have been discussed. In the following, let us summarize the requirements that are related to the server operating systems.

- Support networking and system administration.
- Support application software.
- Support database administration.
- Host Web servers.
- Support remote access.
- Provide directory services.
- Support client-server computing.
- Provide tools for enforcing security measures.
- Support distributed computing.

As mentioned earlier, a server can be either in a stable condition to support other applications or in an unstable condition for students to get hands-on practice on setting up a server, and so forth. Stable servers are used to support courses such as client-server computing, database, application design, and various programming courses. Unstable servers are usually used for courses such as networking and system administration. We will look at the server operating system configuration for both conditions.

Configuration for Courses Using Unstable Servers

To support hands-on practice in courses such as networking and system administration, each student should be given his/her own servers in the lab. On these servers, students have to be system administrators to practice networking and system administration. They will change server configurations from time to time, and the servers may get crashed from time to time. In such a case, we may consider setting up a lab like the one below.

Lab Structure

As an example, let us suppose that there are 16 students enrolled in a network management class. To meet their hands-on practice needs, we decide to install a server operating system on each of the 16 computers. By the past experience, we know that some of the servers need be reinstalled due to errors made by students during hands-on practice. To support the rescue effort, we need a stable server which we call the instructor server, which will be used to hold the image of the original hard drive for those unstable servers. If needed, the image will be reinstalled to a crashed server. The instructor server will also be used to manage the lab network. In the following, let us see how to configure each type of server.

Instructor Server

For the instructor server, you will need to install a server operating system with network management utilities. The instructor server has the following configuration requirements.

*   Name the instructor server as instructor and give a proper password for the lab administrator.
*   Create an account for the instructor and assign the administrator privilege to the instructor.

- Create 16 student accounts and assign each account as a regular user.
- Partition the hard drive into two drives, one for the operating system and one for storing the image of the student servers.
- Configure the network card so that the instructor server can communicate with the student servers.
- Configure domain name server (DNS) so that it can resolve the name for each of the students' computers.
- Configure the gateway so that the computers in the lab can access the Internet or other networks.
- Create a folder that can be shared by the student servers.
- Configure the basic security measures such as the log-on restrictions.
- Configure Dynamic Host Configuration Protocol (DHCP) so that the students' computers can automatically get IP addresses.
- If needed, configure the directory services so that the instructor server becomes a domain controller.
- If needed, configure Network Address Translation (NAT) for the Internet connection.

There may be other configurations needed for a computer lab to meet the hands-on practice requirements; add them if necessary.

Student Server

Since student servers are unstable, the configuration of these computers is basic. The following is a possible configuration.

- Name the computers configured as student servers with names such as student1, student2, to student16.
- Tentatively configure the student servers to receive their IP addresses dynamically from the DHCP server on the instructor's computer.
- If needed, join the student servers to the domain managed by the instructor server.
- Map the shared folder on the instructor server to the local drives so that the student servers can share the files with the instructor server.

The above is the basic configuration. Students will further develop their servers based on the instructions in their lab manual.

Disk Imaging

Disk imaging is a process to clone an entire hard drive from a model computer to other computers. During a semester, the disk images can help recover the crashed computers quickly. In between semesters, these images can be used to quickly reset the computers in a computer lab to the original state to get ready for a new semester. If all the computers in the lab are installed with the Microsoft Windows operating system, the disk imaging can be done with the utility, Remote Installation Services (RIS), included in Microsoft Windows 2003 Server. Otherwise, you may have to purchase the commercial disk imaging products such as Norton Ghost.

When imaging the original hard drive, the system preparation tool (Sysprep) can be used to automate the deployment of the original hard drive to multiple computers. With Sysprep, you can avoid some deployment problems such as the problem that all the imaged computers have the same computer name. To image a hard disk, you may need to perform the following tasks.

- Install all the software on a model computer and configure the model computer to meet all the requirements.
- After the model computer is configured, run the Sysprep tool to prepare the model computer for cloning.
- Image the hard drive of the model computer and save the image to the instructor server.
- Create a bootable installation floppy disk or CD.
- To clone the hard drive of the model computer to other computers, insert the bootable floppy disk or CD to the computer that will receive the image, run the installation file on the floppy disk or CD.
- Install the image to the receiving computers.

Depending on the size of the image, the imaging process may take a few minutes to a few hours. Also, notice that the student computers and the instructor server must be on the same network and all of the student computers must have the same hardware structure.

Virtual Machine

The unstable student servers can also be implemented with virtual machines. If the instructor server has adequate speed and enough memory, you can create virtual machines for each student on it. Each virtual machine has a server operating system installed and configured as mentioned in the student server topic. To create virtual machines, you need to perform the following tasks.

- Install virtual machine software on the instructor server.
- Create virtual machines.
- Install the student server operating system on the first virtual machine.
- If needed, install other software on the first virtual machine.
- Configure a floppy disk drive and CD ROM drive for the first virtual machine.
- Create a virtual network.
- Set up communication between the first virtual machine and a shared folder on the instructor server.
- Run the Sysprep service software.
- Copy the virtual machine to other computers in the lab.

The greatest advantage of a virtual machine is that it simplifies lab maintenance. When a student's virtual machine crashes, all you have to do is to replace the crashed virtual hard drive with the original virtual hard drive which is simply a file saved on the instructor server. Often, it takes only a few minutes to get it done. The same procedure can also be used to prepare the lab for a new semester.

Configuration for Courses Using Stable Servers

Courses such as database management and programming languages need a stable server to host application software. For this type of course, an online computer lab should have stable servers. A stable server is highly secure. Students will not be given an administrator's privilege to the stable server.

Lab Structure

For the lab that depends on a stable server, a highly protected instructor server can get the job done. The instructor server will host all the application software and provide other services for managing the lab. The instructor server needs to be highly reliable so that students can use the lab 24 hours a day, 7 days a week. In such a case, it is better to construct the server on a RAID structure. Some of the application development tasks need the client-server architecture to support hands-on practice. In such a case, the computers in the lab should be divided into several groups and each group of computers will form client-server architecture. In the following, let us discuss the configuration of the stable servers.

Instructor Server

For the instructor server, in addition to a server operating system and network management utilities, you also need to install application software. The configuration of the server operating system is similar to what was mentioned in the previous paragraph. The configuration of each application will depend on the individual application software. To support a highly reliable and high-speed computing environment, you may want to implement the RAID structure on the instructor server. Suppose that you will implement RAID 5 on the instructor server. The following will address some configuration issues on creating a RAID 5 stripe set with parity on three hard drives.

- Verify that a minimum of three hard drives are installed on a server.
- Create at least two dynamic disks.
- Establish the RAID 5 volumes by specifying the same-sized unused volume on each of the hard drives.
- Assign each RAID 5 volume a drive letter and specify a format for the volume.
- Add data to the RAID volumes.

In the above, you are using server operating system utilities to implement a RAID 5 structure. Running a RAID 5 will consume some CPU power on parity calculation. Therefore, running a RAID 5 on the instructor server will not gain as much performance as initially thought. It is better to implement RAID on a different server. Another solution is to implement RAID through hardware.

Group Server

If a client-server architecture is required for teaching and hands-on practice, you will need to install several stable servers in the lab, one for each group. The group servers will host the application software for the groups. The group members will have accounts on each group server. Some of the application development tasks such as Web service development require students to have an administrator's privilege. In such a case, you have to allow students to be administrators. This may lead to a potential risk that students may misconfigure the group servers. However, the risk is smaller when compared with that in a networking class. A better solution is to run the application software on a virtual machine. In such a way, even though the students have the administrator's privilege for the virtual machine, they will not damage the host server if they make mistakes. The disadvantage is that the performance decreases, especially for an online computer lab.

Disk Imaging

For an online computer lab supported by stable servers, you will need to make images of both the client computers and group server computers. The disk images will be saved on the instructor server. Make sure that the instructor server has enough disk space for all the images.

Remote Access

Some of the server operating systems, such as Microsoft Windows Server 2003 and Linux, include virtual private network utilities. In such a case, you may configure the server operating system on the instructor server to set the VPN service for remote access. Or, you can place the VPN service on a separate server to reduce the workload of the instructor server. The computer hosting the VPN should be highly reliable. Once the VPN server is down, the whole computer lab will be down with it. It will have serious impact on multiple courses. The configuration of the VPN server includes the following tasks.

- Configure filtering rules for remote access.
- Configure protocols such as SSH and PPP for security and running TCP/IP traffic.
- Create user accounts on the VPN server.
- To reduce input/output bottlenecks, configure multiple network cards and use one card for each subnet.
- Use a UPS to support daily operations of the VPN server.

If the server operating system is Microsoft Windows 2003, you can also configure Terminal Service for remote access.

Database Server

To support teaching and hands-on practice for database related courses, database management system (DBMS) software should be installed in the computer lab. There are two types of database related courses, database administration and database application development.

For database administration related courses, students will be administrators for a DBMS. In the computer lab, students will work as a database administrator that may perform the following tasks.

- Install and upgrade the DBMS software package on a server.
- Install and manage data storage devices such as external hard drives, optical drives, or tape drives.
- Design and create databases.
- Transfer data in and out of a database.
- Configure a database and network for remote access.
- Deploy database application packages to front-end users.
- Create database user accounts and specify permissions for those users.
- Enforce security measures to protect the database server and sensitive data.
- Backup a database that is currently running and restore a database after the database crashes.
- Optimize database performance by fine tuning the database components.
- Monitor and audit database operations.
- Develop a data warehouse for decision support.
- Set up Web services for database applications.

To support the above hands-on activities, each server in the client-server architecture should have a database server software package installed. The client computers in the lab or in students' homes should have the database client software package installed.

For database application related courses, students will perform the following operations.

- Install and configure database client software on client computers.
- Query database to collect information.
- Develop forms and reports for front-end users.
- Program stored procedures for database application logics.
- Integrate database applications with the supporting database.
- Conduct data analysis such as data mining to identify market trends.

Most of these activities do not require students be an administrator for the server operating system. Therefore, the DBMS software package can be installed on the instructor server.

In general, to install and configure DBMS software, you need work on the following tasks.

- Select a proper version of database server software to install. There may be different versions such as enterprise version or standard version.
- Specify the type of database to create. A database may be used for online transactions or used for data warehousing.
- For remote access, specify the database server's global name.
- Specify locations to store database files. For better performance, the data files, log files, and system files should be stored on different drives.
- Configure the connection string to allow students to log on to the database server from remote locations.

For database administration related courses, the above configuration may be adequate for most cases. Students will learn more about database administration during their hands-on practice. However, to prepare for the database application related courses, more configurations are needed such as those listed below.

- For a database application course, you need to create one or more databases and populate the database with data required by the database application course.
- Create roles and user accounts related to the roles.
- Grant privileges to each role and specify the amount of disk space for each user.
- Develop a maintenance plan including a database backup schedule and a database restoration plan.
- If necessary, implement some measures to improve database performance such as creating a RAID structure.
- Enforce some measures to protect the data such as setting up restrictions to prevent students from accidentally deleting certain rows in a table.

To make sure a database is ready for the class, it is necessary to ask the instructor to test the newly configured database. If necessary, modify the database configuration accordingly.

XML Web Services

XML Web services can be used to support database application courses, e-commerce related courses, and Internet programming related courses. An XML Web service needs a stable environment. It plays an important role in the client-server environment. XML Web services are used to process front-end users' requests and retrieve data from database servers based on the users' requests. The development of an XML Web service project includes two tasks, creating an XML Web service and accessing the XML Web service. The tasks of configuring an XML Web service are summarized below.

- Analyze the teaching and hands-on practice requirements for the XML data, user-defined functions, stored procedures, and utilization of XML data in lab activities.
- Install and test XML Web service development tools. The commonly used Web service tools are IBM WebShare, Java Web Services, Oracle 10g, Microsoft Visual Studio .Net, and SQL Server.
- Configure the services to provide data, functions, and stored procedures to front-end users.
- Deploy the services so that front-end users can use the services from remote locations.

After an XML Web service is configured, students can request the server that is running the Web service to perform certain tasks. When the requests are sent to the server, the Web service executes the requested operations and returns responses back to the clients through the Internet. The XML Web Service allows the students to run highly interactive browser-based applications on the Internet while keeping these students from directly accessing the back-end database; this gives better performance and security.

The Web service depends on the Web server. Therefore, before implementing the Web service, make sure the Web server is functioning properly.

Business Process Automation Server

To take advantage of the power and benefit of Web services, one can use a business process automation server to automate the tasks in a business process including business process design, development, management, and monitoring the collaboration effort in developing a dynamic distributed application system. For an e-commerce related course, a business process automation server can be used to support various kinds of hands-on practice. If the course is about application development, the business process automation server needs a stable computing environment and requires students to be system administrators. Hence, the server should not be installed on the instructor server. The following are the tasks of configuring a business process automation server.

- Design an integrated solution for a business process and make an implementation plan.
- Create a server cluster for the installation of the business process automation server.
- Install and configure Web server software, Web service development tools, and database client software.
- Create user accounts and place these users in a group.
- Install the business process automation server and specify the development environment to meet the hands-on practice requirements.
- Configure network balancing on multiple nodes.
- Configure runtime to execute the operation of a computer program.
- Creating the business process automation server hosting environment.
- Develop schemas, pipelines, transactions, and error handling for process automation.

To be able to create solutions with a business process automation server, in addition to the global system administrator account, students also need the database administrator account.

E-Mail Server

Some technology-based courses may require hands-on practice on developing e-mail servers. If the hands-on practice is about e-mail server installation and administration, e-mail server software should be installed on the group servers. The following are tasks for e-mail server installation and configuration.

- For each server in the client-server architecture created in the computer lab, install an e-mail server.
- Create user accounts and specify the authentication method.
- Configure the POP3 service which is a network protocol for delivering e-mails across the Internet.

- Enable the sponsored projects accountant (SPA) to have secure communication between the e-mail server and the e-mail client.
- Configure the simple mail transfer protocol (SMTP), which is a protocol used for receiving and sending e-mails.

After the configuration is done, you can use a client account to verify the set up. If you are not satisfied, modify the configuration accordingly.

Customer Relationship Management (CRM) Server

An e-commerce course may require the CRM software package for its hands-on practice. CRM is used to help an enterprise manage customer relationships. CRM may include components such as a database or directory service to hold customer information, e-mail system for communication with customers, Web services for providing services, application software such as Microsoft Office. The tasks of CRM configuration are summarized in the following:

- Check the hardware and software requirements for the deployment of CRM.
- Install the CRM server, CRM client, and e-mail router.
- Initialize the system administrator account.
- Create CRM business units, users, and teams.
- Configure CRM security roles, privileges, and access levels.
- Customize application components such as a database, XML services, e-mail, and reporting services.
- Migrate customer data to the database.

If the course requires students to learn how to install and configure CRM, the lab administrator should skip the above steps and let students complete all the configuration tasks.

Web Conferencing Server

To support online live lectures and Web conferencing, you can install Web conferencing server software on the instructor server. A Web conferencing server allows a group of students to work together. It can also be used to schedule meetings, integrate VoIP audio broadcasting, control presentations, and upload multimedia files. In general, you may need to perform the following tasks to create a Web conferencing server.

- Check the hardware and software requirements.
- Install the Web conferencing server software.
- Create user accounts and specify permissions.
- Configure the Web conferencing environment and services.
- Determine the trusted computers and configure collaborations.
- Check audio and video components.
- Configure remote access to the Web conferencing server.

Once a Web conferencing server is set up, it is ready for users to schedule meetings and presentations.

So far, we have discussed the general installation and configuration of several special purpose servers. It is not necessary to install these special purpose servers all at once. It depends on the needs of the technology-based courses.

Conclusion

In this chapter, topics related to server configuration and implementation issues have been further investigated. The configurations for low-end, midrange, and high-end servers have been discussed in this chapter. We have discussed the configurations of server hardware, uninterruptible power supplies, server operating systems, and some special purpose servers. To properly configure the server-side hardware, we have investigated the requirements for the server hardware by small, medium, and large computer labs. For small, medium, and large computer labs, general configuration guidelines have been given for a group of stand-alone servers, rack systems, and blade systems. For medium and large labs, a rack system or a blade system has been recommended for saving space and for better performance.

To improve reliability and performance, this chapter has introduced the RAID systems. We have discussed different types of RAID systems and the configuration information about these RAID systems. For a large computer lab, this chapter has also provided configuration information of the cluster and grid computing environments. To avoid server system shutdown during a power outage, this chapter has looked into various uninterruptible power supplies. We have discussed two types of UPSs, the standby UPS and generator-based online UPS.

This chapter has also dealt with the configurations of server operating systems. The configurations of unstable and stable server operating systems have been discussed in this chapter. For remote access, we also looked at the configuration of a VPN server in this chapter. The final portion of this chapter provided the configuration information of various special purpose servers such as XML Web services, database servers, business process automation servers, and e-mail servers.

This chapter has accomplished the server related tasks in an online computer lab development process. After the servers are installed and properly configured, our next task is to install and configure the network equipments and network management software. We will cover these topics in the next chapter.

References

APC. (2006). Select your protection needs. Retrieved April 5, 2007, from http://www.apcc.com/tools/ups_selector/index.cfm?lid=Go%20to%20the%20UPS%20Selector

Ayala, G., & Paredes, R. (2003). Learner model servers: Personalization of Web based educational applications based on digital collections. In D. Lassner & C. McNaught (Eds.), *Proceedings of World Conference on Educational Multimedia, Hypermedia and Telecommunications 2003* (pp. 432-435). Chesapeake, VA: AACE.

Boer, A. (2002). A delivery server infrastructure for multimedia distance learning services. In P. Kommers & G. Richards (Eds.), *Proceedings of the World Conference on Educational Multimedia, Hypermedia and Telecommunications 2002* (pp. 178-179). Chesapeake, VA: AACE.

Buchanan, W., & Wilson, A. (2000). *Advanced PC architecture*. Indianapolis, IN: Addison-Wesley Professional.

Chapa, D. A., & Radu, P. (2006). *The blade server guide: Cutting the cost of enterprise computing*. Indianapolis, IN: Wiley.

Englander, I. (2003). *The architecture of computer hardware and systems software: An information technology approach*. Indianapolis, IN: Wiley.

Fansler, K. W., & Riegle, R. P. (2004). A model of online instructional design analytics. In *Proceedings of the 20th Annual Conference on Distance Teaching and Learning*. Retrieved April 5, 2007, http://www.uwex.edu/disted/conference/Resource_library/proceedings/04_1069.pdf

Hunt, C. (2002). *Linux network servers*. Alameda, CA: Sybex.

Intel Corporation. (2006). Itanium® 2-based systems. Retrieved April 5, 2007, from http://www.intel.com/business/bss/products/server/itanium2/index.htm?ppc_cid=ggl|64bit_itanium|k16AE|s

Joseph, J., & Fellenstein, C. (2003). *Grid computing* (On Demand Series). Indianapolis, IN: IBM Press.

Lowe, S. (2001). Building out a rack-based server farm. TechRepublic. Retrieved April 5, 2007, from http://techrepublic.com.com/5100-1035-1040754.html

Minasi, M., Anderson, C., Beverridge, M., Callahan, C. A., & Justice, L. (2003). *Mastering Windows Server 2003*. Alameda, CA: Sybex.

Patterson, D. A., Gibson, G., & Katz, R. (1989). A case for redundant arrays of inexpensive disks (RAID). In *Proceedings of the International Conference on Management of Data* (pp. 109-116).

Poniatowski, M. (2005). *HP-UX 11i Version 2 system administration: HP Integrity and HP 9000 servers.* Upper Saddle River, NJ: Prentice Hall PTR.

SPARC International. (1994). *SPARC architecture manual Version 9.* Upper Saddle River, NJ: Prentice Hall PTR.

Vetter, S., Pruett, C., & Strictland, K. (2004). IBM eServer Certification study guide: pSeries AIX System Support. IBM Redbooks. Retrieved April 5, 2007, from http://www.redbooks.ibm.com/abstracts/sg246199.html?Open

Vetter, S., et al. (2005). Advanced POWER virtualization on IBM eServer p5 servers: Architecture and performance considerations. IBM Redbooks. Retrieved April 5, 2007, from http://www.redbooks.ibm.com/abstracts/sg245768.html?Open

Chapter VI

Network Development for Online Computer Labs

Introduction

In the previous chapter, we have discussed the issues relate to the server-side configurations. After servers are developed, our next task is to develop networks that will connect the servers to the client computers. Networks are also used to teach networking related courses. In these courses, students work on network configuration, network management, network security, and Web related tasks. These types of trainings are necessary to meet the requirements from the e-commerce industry. Our students need a network administrator's account for practice. Being a network administrator gives students a good opportunity to gain network problem solving skills.

The configuration of a network will depend on the needs and requirements of teaching and hands-on practice. In this chapter, we will discuss issues related to the implementation of networks for each type of online computer lab model. We will start with the peer-to-peer lab model in which computers are connected through a switch or several crossover network cables.

Next, we will discuss network implementation issues for a small computer lab in which computers are connected through multiple switches and at least one router. Based on the teaching and hands-on practice requirements, the computers in the lab can be configured to join a workgroup or join a domain.

For a medium computer lab, multiple networks will be constructed. Computers in these networks need to communicate with one another. For this type of lab, more network equipment is needed and the configuration is also more complicated. Multiple switches and routers will be used to construct the networks. These computers may be configured to join one of the domains depending on the requirements of teaching and hands-on practice.

In this chapter, we will also discuss the network implementation issues of large computer labs. For a large computer lab, multiple medium online computer labs are connected through a wide area network (WAN) or through the Internet. For this type of network, the network developing process often involves outside companies such as a telephone company and a technology consulting company.

In this chapter, we will specify, in detail, what network equipment to use and how computers communicate with each other within a network and between different types of networks. An online computer lab involves a number of network technologies and equipment. In this chapter, we will investigate how network equipment is used in a network infrastructure and discuss network topologies, network deployment, and management techniques. The discussion will provide some of the up-to-date information in the fields such as wireless LANs (WLANs), VoIP, content networking, and storage networking. We will have a brief discussion about these equipment and technologies, and how they can be used to meet the teaching and hands-on practice requirements.

Lastly, this chapter will present a case study about the network configuration of a medium online computer lab. In this case study, all the servers developed in the previous chapter will be linked together and configured to be ready for an online technology-based course.

Background

Each of our computer lab models requires a network to allow computers to communicate with each other. To meet the requirement, we need to properly construct a network computing environment for an online computer lab. Depending on the operating system installed on the network server, a network can be categorized as a Windows network, a Linux network, or a UNIX network.

For a Windows network, Windows Server 2003 and Windows Server 2000 are the main network operating systems Hunt and Bragg (2005). These network servers provide several network management tools and support various network protocols such as TCP/IP and so on. A network class can also use network simulators. A network simulator is great for hands-on practice on a network. It is software that can be installed on a student's home computer. Windows network simulators can be used for network implementation, management, and maintenance, and can be used for professional training (Chellis & Sheltz, 2005).

The Linux operating system can also be used to manage networks (Bautts, Dawson, & Purdy, 2005). With the Linux operating system, one can configure the network hardware and software such as Linux kernel, Linux network architecture, security measures, routing configuration, Bluetooth, and network protocols such as TCP/IP, UDP, PPPoE (Wehrle, Pahlke, Ritter, Muller, & Bechler, 2004). It can also be used to manage the Internet and Wireless network.

UNIX operating systems are often used to support an enterprise-level network structure which requires a high level of security, availability, and reliability. A UNIX network is often supported by a major network equipment company such as Cisco (Schmied, 2004). Network hardware includes routers, network interface cards, switches, and bridges. An enterprise-level network design and implementation process involves all these hardware equipment (Teare & Paquet, 2005). The network in a large computer lab is similar to an enterprise-level network. In addition to the network configuration tasks in developing small and medium online computer labs, the tasks in developing a large computer lab often include configuring Network Address Translation (NAT), managing bandwidth, setting routing policies, and implementing multicast architectures for a UNIX computing environment.

As one of the important subjects, networking is related to many courses in the computer science and information systems curriculum, from the entry level (Comer, 2004) to the more advanced level (Hassan & Jain, 2003). Networking related courses are developed for teaching network theories and concepts. An introductory networking class covers topics such as an overview of networking. For students who have no previous experience in networking, it explains how the network technology works. A more advanced networking class often teaches network performance analyses, network algorithms, and network protocols.

Stockman (2003) reports the use of virtual system technologies to create virtual networks which improve both the lecture part and the lab component of networking classes. In the classroom, by using such a system, a faculty member can remotely access the computer in his/her office. As mentioned earlier, Correia and Watson (2006) also discuss the use of the virtual network technology in a computer lab for teaching networking courses on campus. The virtual network technology is easy to maintain and reduces the cost on lab hardware. Another significant development is using the wireless network technology. Varvel and Harnisch (2001) report the implementation of a wireless laptop computer lab that provide students with more mobility in learning and give instructors more choice in teaching. The networks in these studies are local networks on campus. We need to extend the accessibility so that students can access lab facilities from anywhere and at anytime for online technology-based courses. Next, we will further look into the issues on the configuration of networks for different online computer lab models.

Peer-to-Peer Computer Lab Network

The network for a peer-to-peer computer lab is the easiest one to be implemented. You may implement such a network in one of the following ways.

Using Crossover Cable

You can connect each pair of computers with a crossover cable. The number of computers that can be connected by this type of network is very limited. As the number of computers in the network increases, the number of cables to be used will increase rapidly. Also, it is very tedious to configure the network interface cards. Realistically, it is only good for a group of two computers.

Using Switch

A better way to implement a peer-to-peer computer lab is to use a low cost switch. The computers are connected to the switch. Through the switch, the computers can communicate with each other. There are several ways to allow students to remotely access these computers. The easiest way is to configure each of the lab computers with a static IP address. Sometimes, network managers are reluctant to give out static IP addresses due to the fact that it is a bit risky to expose the computers to the Internet without good protection. Also, sometimes there may be not enough static IP addresses to give. In such a case, you can configure the computers to accept the dynamic IP address from a dynamic host configuration protocol (DHCP) server on a university's network. Make sure to ask the network manager to configure the DHCP server to bind the DHCP IP address to a lab computer's hardware address.

Using DSL Router

If security is a great concern, the network manager may not want to let any of the lab computers be part of the university's network. A better solution is to use a low cost DSL router to separate the lab network from the university's network. For example, each of the computers in the lab is connected to the router and is configured to take a dynamic address from the DSL router. These dynamic IP addresses will be assigned by the DSL router. Make sure to configure the DSL router to assign a computer with the same dynamic IP address so that the students can remotely access the same computer through network address translation (NAT) built inside the DSL router. The DSL router can take an IP address from the university's network as its external IP address. If security is really an issue, you may have to get the DSL connection directly from an ISP provider without touching the university's network.

Using Server

If the computers in a peer-to-peer computer lab are installed with a server operating system, you can configure one of them to function just like a DSL router. You need to have two network interface cards installed on the server computer used as a router. One of the network interface cards will be used to connect to the university's network and the other one will be used to connect your lab. Windows and Linux operating systems are often configured to perform some of the tasks done by a router. To connect the computers in the lab, an additional switch or hub is needed. Through the switch or hub, the lab computers will be able to communicate with the server and with each other. You also need to configure the DHCP and NAT for the server computer to assign dynamic IP addresses and to allow Internet access. Again, make sure that a computer in the lab will get the same dynamic IP address every time the computer is rebooted.

Small Computer Lab Network

The network for a small computer lab includes equipment such as servers, switches, and routers. Several subnetworks (or subnets) will be constructed for group activities. Routers or bridges will be used to connect each subnet so that computers from different subnets can communicate with each other. In each subnet, the computers are connected through a switch like the network in a peer-to-peer computer lab. Depending on the requirements of teaching and hands-on practice, a small computer lab may have a few variations. In the following, let us take a closer look at some of the variations.

Using DSL Router

In this type of lab, there is no requirement for each group to have a different network ID from another group. In such a case, the computers in each group will be linked to a switch or hub. Each group switch will be linked to a DSL router through the uplink port. In each group, one of the computers can be installed with special purpose server software and the other computers can be configured as client computers. The advantage of this kind of small lab is that it is easy to implement and maintain. It is a good platform for a small client-server computing environment. The disadvantage is that all the servers are in the same network. This does not meet the requirements of some networking related hands-on practice where students are required to develop multiple networks.

Using Single Router

If the DSL router is replaced by a more sophisticated router which can forward data from one network to another network with different configurations, then we may be able to let the computers in one group communicate with the computers in another group which has a different network ID. To handle multiple Ethernet networks with different network IDs, the router should be equipped with multiple Ethernet interfaces. In each group, one of the computers can be the primary server. Depending on the hands-on practice requirements, the other computers in the same group can be the secondary servers, special purpose servers, or client computers in the client-server architecture. Computers in each group have their own network ID and can be configured to belong to the same workgroup. The server in each group can be configured to work as a DNS server, a Web server, a file server, and a DHCP server.

Using Single Server

The functions of a router can also be implemented by a computer if the proper protocol is installed. An easy way to do this is to install a server operating system on the computer and then use the tools provided by the server operating system to forward network packets from one network to another network. You can also route the network packages by using other routing software. If multiple networks are involved, you need to install multiple network interface cards on the server. The server can also be configured as a DNS, Web, file, or DHCP server.

Figure 6-1. Networks in a small computer lab

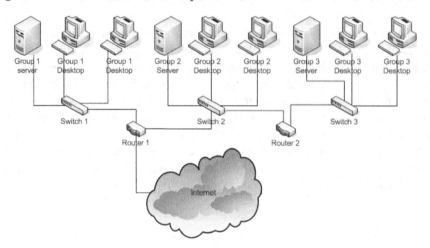

Using Multiple Routers

Multiple group networks can be connected with multiple routers or bridges. By doing so, you can reduce bottlenecks and increase reliability. You may link two group networks with a router that has two network interfaces. Figure 6-1 is a multiple-router network designed for a small computer lab. In Figure 6-1, there are nine computers on the network. These computers are divided into three groups. The computers in each group are connected through a switch. Each pair of switches is linked through a router.

In Figure 6-1, if Group 2 and Group 3 have the same type of network, then Router 2 can be replaced by a bridge which costs less than a router.

Medium Computer Lab Network

In a medium computer lab, networks can grow even more complex. Additional network devices will be used. These networks will be used to support multiple technology-based courses; each course may have its own requirements for the networks. Directory services can be implemented on the multiple networks for students to learn about domains and organization units and to perform network management tasks. Multiple routers or bridges will be used to connect the networks in the lab. For courses that need a network environment to support the client-server architecture, you need to build a stable network to support hands-on activities. Often, the courses such as database management, application development, and programming can share the same stable network. On the other hand, a course such as networking requires students to develop their own networks. The configuration of a network is changed from time to time. It is a good idea not to assign any course that requires a stable client-server computing environment to an unstable network. Therefore, we should separate the unstable network from the stable network. In the following, we will go through some of the networks designed for different courses.

Network for Teaching Application Development

Application development related courses require a stable network environment to support the client-server architecture. The networks described for a small computer lab can get the job done in here too. The server computer in each group will host the special purpose server software. The application client software will be installed on the client computers. After the networks are built, multiple courses in programming or database systems can share the same network environment.

To better administrate a client-server computing environment, directory services can be used to ease the network management tasks and improve network security. Some application software, such as Microsoft SQL Server, can be installed on a stand-alone computer or installed on a member server in a domain. To use the application software like Microsoft SQL Server in a client-server environment, it is better to install the server component of the software on a member server in a domain and configure all the client computers to join the domain. The following are the tasks for configuring a domain for a medium computer lab.

* Plan for installing directory services by specifying the domain structure, IP addresses, subnet masks, default gateway, and so on.
* Physically wire all the computers that belong to the client-server architecture.
* Install a server operating system on the dedicated server computer.
* Configure the DNS server, DHCP server, and network adapters.
* Install directory services software. Configure the directory services to either join an existing domain or to create a new domain.
* Design the directory services to support network management and the application of group policies.
* Configure the directory services by creating users, groups, and roles.
* Enforce the security measures to protect the networks in the lab.
* Create a Web site.
* If needed, create more domains or other directory service objects.
* Configure trust links among these domains.

- Develop and implement a directory service replication plan.
- Develop and implement a directory service data recovery plan.
- Develop and implement a data recovery plan.
- Implement the group policies for the desktop applications.
- Install application software for client-server computing.

For most of the client-server computing related courses, a single domain should be adequate. On the other hand, for network management or system administration related courses, multiple domains are often created to meet the requirements of teaching and hands-on practice.

Network for Teaching System Administration and Network Management

Courses related to system administration or network management require a stable network structure in which the connection of network hardware remains unchanged. However, some of the network configurations will be changed based on the requirements of teaching and hands-on practice. It is difficult to host client-server architecture on this type of network. After a network is reconfigured, it may prevent client users to log on to the server. If an application development course needs to share the computer lab with a network management course, you may consider creating two separate networks to meet the requirements from each course.

For a system administration or network management course, the hands-on practice may require each group to have a decentralized network administrative structure. It needs the networks to have distinct sets of system and network administrators, and group policies. To achieve this goal, a possible network setup is that the computers in a group are configured to join a subdomain. One of the computers in each group serves as a domain controller. These are the tasks to create a subdomain for each group:

- Design and implement a root domain on the instructor server.
- Design a subdomain for a group of computers.
- Implement the subdomains, one for each group.
- Implement trust relationships between the subdomains and the root domain.
- Create administrator accounts for each subdomain and the root domain.
- If necessary, create some organization units in each domain.

- Assign group policies to an organization unit or a subdomain.
- Configure the computers in each group to join the group's subdomain.

Even though the configurations of domain and organization units can be changed, this type of network requires a stable hardware structure. After the routers and network interfaces have been installed and configured, the structure should remain the same through out the semester.

Network for Teaching Network Infrastructure

Some other courses such as network infrastructure related courses may require you to reconstruct a network from the ground up. In a network infrastructure class, students often need to create multiple areas in an autonomous system or even multiple autonomous systems. Students will learn how to configure some routers with Interior Gateway Protocol (IGP) and some routers with Exterior Gateway Protocol (EGP). The network for each group will be redesigned during a semester. Hardware devices can be added or removed for each network design. Due to its instability, it is difficult to use this type of network to support other technology-based courses. The network of this type should be separated from other networks. Often, this kind of network does not have to be fully constructed. For an online class, you may need to prewire all the equipment and install server operating systems or other files used for network configurations. Later, students will perform tasks such as designing and configuring the network and decide which device use in the network. Sometime, you may need a lab assistant to rewire the network before a new lab session.

Network for Teaching Network Security

Another type of network that cannot be shared by other courses is the one that supports network security related courses. For this kind of network, you need a stable network to support activities such as enforcing security measures. As a hands-on exercise, students need to configure filters or firewalls to block certain protocols, networks, computers, or users for network protection. These activities can make a network unavailable to other users. Therefore, a network security course should have its own network. The network for teaching network security usually has multiple subnetworks so that students in each group can configure their own subnet for network security.

To sum up, we may have a possible medium computer lab shown in Figure 6-2. This lab can be used to support courses such as application development, networking, system administration, security, database, and so on.

Figure 6-2. Networks in a medium computer lab

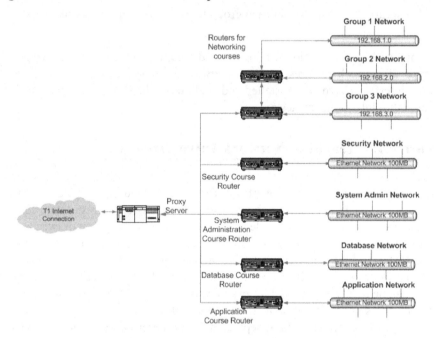

The networks, such as the network for security or system administration, can be further divided into subnetworks for group activities. Some technology courses may require each student to have his/her own network which contains at least two computers. In such a case, the network can cost a lot and can be very complicated. To overcome this difficulty, you may consider the virtual network technology (Correia & Watson, 2006).

Virtual Networks

As we have seen, you may be required to construct multiple networks in a medium computer lab to support different types of networks. If it is needed, you may have to link these networks designed for different classes with routers to allow communication among different networks. Many additional network equipment needs to be added, making the cost of networks increase dramatically. It will also make the lab maintenance more complex. To reduce cost and simplify maintenance, one of the solutions is to use network simulation software or virtual networks.

Network simulation software can be used to design and implement networks at the enterprise level. It does not take a lot of computer lab resources and is very flexible to use. For the following hands-on practice tasks, you may consider using network simulation software.

- A network area may include large numbers of computers and network equipment. In such a case, it is better to use network simulation software to design and implement multiple network areas that are interconnected in an autonomous system.

- An autonomous system may contain computers and network equipment for a global enterprise. It is difficult for a university to physically implement multiple autonomous systems. In such a case, network simulation software can be used to design and implement global autonomous systems.

- As mentioned earlier, some of the network courses require students to redesign the network infrastructures. They may require students to rewire the computers and network equipment; this is difficult for an online class. Network simulation software is a convenient tool that can be used to redesign previous existing networks or autonomous systems.

- When a network course requires students to develop a wide area network (WAN) or an Internet related project, using network simulation software is a better solution, due to the large scale of the project.

- A course of network design often requires students to analyze the network for an entire enterprise. Network simulation software provides graphical utilities to make students see the big picture of the entire network topology. It will help students to better understand an enterprise level network system.

- Some network related courses are more focused on configuration issues and less concerned about hardware connections. In such a case, network simulation software is a perfect solution.

To safely carry out hands-on practice online with IT products on the current market, you may consider using the virtual machine technology. On a physical server, multiple virtual networks can be created. That will reduce the number of computers and network equipment. Once students log on to a virtual machine created in a virtual network, they can feel like sitting in front of a physical machine. It is one of the advantages of a virtual network, which can help students gain experience on using products of current technologies. Virtual networks have some other advantages too. For example, by using a virtual network, it is much easier to maintain daily operations. The tasks for upgrading and reconstructing computer labs for a new semester are much easier to carry out. Working on a virtual network is also much safer than working on a physical network. Students do not need to have the administrator's privilege on the physical host server. They can use a regular user account to log on to the physical host server and then use an administrator account to log on to a virtual machine. Also, the cost of a virtual network is significantly less than the cost of a physical network with a similar configuration. One disadvantage of a virtual network is its performance. When accessing multiple virtual machines created on a physical server through the Internet, students can experience a significant slow down in performance. Virtual machines have some other minor problems such as being incompatible with some hardware such as certain brand floppy disks and power management utilities such as hibernation. The following are some tasks of setting up virtual networks for hands-on practice.

- Install and configure a physical server which allows Internet access and has enough memory and hard drive storage space.

- Create student accounts on the physical server to allow students to access the server through the Internet.

- Install virtual server software and create multiple virtual machines with the virtual server. These virtual machines will be used as servers to support various technology-based courses.

- For each virtual machine, configure hard drives, a floppy drive, and a CD-ROM drive.

- Configure the network between the physical server and virtual machines, so that the physical server can communicate with the virtual machines.

- Create a shared folder on the physical server so that the virtual machines can store or exchange files in the shared folder.

- Install server operating systems on selected virtual machines which will be used as servers for each virtual network.

- If necessary, configure a redundant array of independent (or inexpensive) disks (RAID) structure to improve reliability and performance.

- Configure DNS, DHCP, directory services, or other network services depending on the requirements of teaching and hands-on practice.

- Create multiple independent virtual networks used to support various technology-based courses.

- Configure NAT or a network bridge on a virtual server of the virtual network which can communicate with the physical server.

- If needed, create some virtual client machines in each virtual network based on the lab requirements of technology-based courses.

- Configure hard drives, floppy drives, and CD-ROM drives for virtual client machines.

- Install client operating systems and application software on virtual client machines.

- Configure the network cards on virtual client machines so that the virtual client machines can communicate with the virtual server.

- Install the System Preparation Tool software on a virtual client machine and make a copy of the hard drive of the virtual client machine.

- Attach a copy of the virtual hard drive to other virtual client machines. You can also make an image of the virtual client machine with disk image software. Then, image the virtual client machine to other virtual client machines all at once.

- If needed, configure a virtual server machine to work as a router. This kind of router allows virtual machines for different virtual networks to communicate with one another.

A virtual network is particularly useful in supporting hands-on practice on client-server architecture. Each student can have his/her virtual network which connects two or more virtual machines. Students can be the administrator of all the virtual machines on his/her virtual network. This has overcome many obstacles due to insufficient privilege. Through the virtual client-server architecture, students can set up a back-end database server, implement Web services and business logics on an application server, and develop forms and reports on a virtual client machine. On a virtual network, an application server will communicate with the database server to retrieve or store data. On the virtual network, the applications created on the virtual client machine will consume the data provided by the application server.

A virtual network is also a good platform to support network security related courses. Configuring filters or firewalls for an individual virtual network will not prevent students from logging on to other virtual networks and the physical server. By monitoring a virtual network, students can observe the effects of enforcing security measures.

Another good use of virtual networks is to support network management courses. On each virtual network, students can practice network configurations, create domains and organization units, define and implement group policies, create user accounts with different permissions, enforce security measures, configure network services, and carry out performance tuning tasks.

Virtual networks are suitable for teaching networking related courses. Students will learn how to configure network interfaces and how to communicate with other virtual machines on the same network. On a virtual network, students can experiment various network protocols and monitor network traffic. By converting a virtual server machine to a router, students will learn a lot about a router. They will learn how to install and configure routing protocols and multiple network interfaces. Through the hands-on practice on a router project, students will see how packets are routed from one network to other networks.

Virtual networks can also be used to support the Internet computing related courses. For the development of a Web server, students are often required to have the administrator's privilege. It is not a problem for them to be an administrator on the virtual network. If NAT is preconfigured, students will be able communicate with the Web server in an online computer lab from their home computers thorough the Internet.

For each technology-based course, there could be multiple virtual machines attached to various virtual networks. To reduce the workload for daily maintenance, one should make at least one backup copy of each type of virtual machine and store it in a safe place. These backups can be used to repair crashed virtual machines. They can also be used to reconstruct the computer lab before a semester starts.

Large Computer Lab Network

A large computer lab network is often constructed on a wide area network (WAN) or the Internet. In a WAN, the data are private and the network generally requires T1 service. The data transmitted over the Internet are not private in general. To connect to the Internet, you need the services from an Internet service provider (ISP). The university's computer service department needs to work with its telephone company and the ISP to establish connections among campuses. When developing networks for a large computer lab, four areas need to be taken care of.

- **Network connectivity:** Make sure that the networks on different campuses can communicate with each other through the Internet or by a WAN.
- **Network security:** Protect the university's internal networks and the data transmitted over the Internet.
- **Network performance:** Detect bottlenecks in data transmission and prevent them from happening again. The networks should be tuned to gain better performance. The network load balancing services should also be implemented.
- **Network reliability:** Minimize the impact of Internet outage, telephone service outage, and power outage and make sure that data can be transmitted to users who are qualified to receive the data.

To allow the internal networks to communicate through the Internet, one needs to determine the services and devices to use in a large computer lab. The following are some general configuration ideas to accomplish the tasks listed in the above areas. The requirements for network connectivity, security, reliability, and performance can be used as a guideline for a large computer lab network project.

Network Connectivity

To communicate with the network on another campus, you need to install and configure the remote access devices. The descriptions related to connectivity issues are given in the following subsections.

Connecting Multiple Campuses

The first task is to determine what type of connection to use to connect multiple campuses. You may either use a private direct connection or through the public Internet connection. Private direct connections are often provided by phone companies. It is a highly secure connection. The disadvantage is that it also has a high price tag. For computer labs, it is not worth it to pay a high price to protect information generated by lab activities. Therefore, most of the computer labs use the public Internet connection.

Remote Access Methods

To properly configure the network connections, you also need to find out what types of networks to access and what types of remote access tools are available on the client side. The common remote access service is T1 used by universities to transmit data from one campus to another campus. An analog modem, DSL, and cable are often used to transmit data from students' homes to computer labs. An analog modem is too slow for the hands-on practice using GUI tools. If the use of GUI tools is required, try to avoid the analog modem connection. However, if the analog modem is the only way of remote access for a large number of students in a class, configure the analog modem connection and redesign the lab activities to avoid using GUI tools.

Border Routing

For internal networks to communicate through the Internet, the following services and devices need to be installed and configured. The routing to the Internet can be done through a static route from a campus to an ISP, or it can be accomplished through an exterior routing protocol such as a border gateway protocol (BGP). The BGP gives a much better flexibility. When one route of communication is stopped due to a network problem, the BGP will try to find an alternative route; this will significantly improve network reliability.

Network Security

Security is a great concern for a large computer lab network since data transmitted over the Internet can be potentially unsafe. Some of the security related topics are discussed below.

Data Protection on the Internet

To securely transfer data from one campus to another campus through the Internet, one often uses the virtual private network (VPN) services. When configuring a VPN, you can set VPN access policies and specify the encryption mechanism used by a VPN service.

Firewall Configuration

To protect the university's networks, a firewall should be installed and properly configured according to the network security policies. If it is necessary, isolate the firewall server from the internal network to prevent users on the firewall server from directly accessing the internal networks. Install the DNS server (software) on a different server from the firewall server unless the firewall serves as an external DNS server. Also, try not to install Internet related services such as hypertext transfer protocol (HTTP) and file transfer protocol (FTP) on the firewall server.

Service Distribution

Using multiple servers with each performing a specific task can also improve security. Physically separating devices always provides more security than logically separating the devices. In a distributed environment, users are given different privileges for each device based on their operation roles. In such a way, one person who has the administrator's privilege on one service may not have the administrator's privilege on another service. This can prevent a person from causing significant damage to a large network. In addition, when attackers paralyze one server, the other servers will still be available to support the lab operations.

Network Performance

There are many ways to improve network performance for a large computer lab. It can be done by changing the network infrastructure, selecting different hardware and software, and properly configuring the network system. The following describes some of the commonly used methods for improving network performance.

Grid Computing System

For better performance, you should try to avoid using a single device, such as a network interface or a server, to carry out all the workload. Due to the fact that a large computer lab is often across multiple campuses, the servers on different campuses can share the workload. The service and workload should be distributed among the campuses.

Network Equipment

When constructing the networks for a computer lab, select faster hardware equipment. For example, choose a network interface card that has a gigabyte-transmission rate instead of a 100Mb rate and choose a wireless network card based on the 802.11g standard instead of one based on the 802.11b standard. Select a server with more random access memory (RAM) and faster processing speed. The Internet transmission rate can also make a big difference. However, adding new hardware is often limited by budget.

Client-Server Architecture

The client-server architecture can be used to improve server performance. If most business logics are computed on the middle-tier application server and only necessary data are retrieved from the database server which is used to store and manage data, the computation load will be drastically reduced on the database server. In this way, you can achieve much better performance.

Multiple Network Interface Cards

To reduce the I/O bottleneck, a server can have multiple network interface cards. For example, one of the network interface cards can be used for the Internet traffic, another one can be used for the internal network traffic, and the third one can be used for network storage.

Bandwidth Usage Priority

Another way to efficiently use a network is to configure the priority of bandwidth usage. In such a way, you will make sure that more important applications get a higher priority to use a large portion of the available bandwidth.

Data Compression

Data can be compressed before they are transferred over the Internet. The compressed data take much smaller bandwidth.

Network Traffic Reduction

During a live lecture, high density life images can significantly slow down the network performance. If lower density images are satisfactory, use the low density images instead or simply keep the audio part only. Also, reduce group e-mail which takes a large portion of the network bandwidth, especially graphical files. If a computer on the network is infected with a virus, it will generate a lot of junk traffic on the network. To stop junk traffic, make sure that antivirus software and spam filters are installed and are functioning properly.

Network Reliability

To make an online computer lab available to students 24 hours a day, 7 days a week, we need to construct a reliable network system. Some structures and services can be used to make a network more available.

Grid Computing System

Developing a grid computing environment is one way to improve network reliability. For better reliability, you should distribute the services through different devices. If one of them fails, it will not impact the support of other services.

Redundancy

Reliability can also be improved by building server redundancy. Redundancy can be implemented by creating a mirror of servers, or by constructing a RAID system.

For high reliability, you can also consider running two identical sets of network devices at the same time, or configuring the second set of network devices as hot standby. Although running redundant servers will increase the cost, the reliability is significantly improved, especially when the servers are distributed across multiple campuses. In this way, the failure of a single device will not cause the entire network to shut down. Network cards can also be teamed up to form redundancy. Each server can have two network cards and each card is connected to different network devices such as switches or hubs. When a network device fails, the communication with other computers can continue through the surviving network devices.

In the above, some of the configuration issues have been discussed. As mentioned earlier, a large computer network is often a team project that involves telephone companies, Internet service providers, and consulting companies. The network should be designed and developed based on decisions made by the team. When implementing a large computer network, we should also consider the needs for future growth. The network should be constructed to allow future expansion.

Case Study of Network Development

To illustrate the development of a network for a computer lab, let us develop a network for a medium online computer lab. This lab will be used to support teaching and hands-on practice for the courses such as networking, network management, application development, and database systems. To see how to design a lab for a specific course, let us look at a case study. In this case study, we can give more specific information about the hardware and software used in the lab. The case study can be used to demonstrate the selection and configuration of hardware and software that are covered in this chapter and in previous chapters. We will first briefly go over the lab manual for each course, and then specify the hardware and software to meet the requirements of hands-on practice. Lastly, we will configure the network for each course.

Hands-On Practice Requirements

For different courses, the requirements for hands-on practice are different. The following briefly summarizes hands-on activities included in the lab manual for each technology-based course mentioned above.

Networking

A networking course may require students to get hands-on practice on the configurations of server operating systems, network equipment such as network interface cards and routers, and client operating systems. Students are also required to configure network services such as DNS, DHCP, NAT, VPN, and FTP. The lab manual may also require students to install and configure special purpose servers such as Web servers and file servers. In the online lab, students will also learn how to monitor and audit network traffic. Some networking courses may also require students to design and implement an enterprise level network infrastructure.

Network Management

In this kind of course, the hands-on practice will require students to create, configure, and replicate a directory service for network management. Students will manage networks at the enterprise level. The lab activities include constructing network management components such as domains, subdomains, organization units, trust relationships, users, groups, group policies, computers, and so on. Students will learn how to configure access control, network auditing, domain update, and information replication. In the online lab, students will perform some network administration tasks such as directory service maintenance, network troubleshooting, network failure recovery, and file distribution. Protecting the network is another task that can be carried out in a network management course. Students will learn how to control remote access, use smart cards, and configure network authentication.

Database System

This is a course that requires students to develop skills in developing a database system. The hands-on practice of this course requires students to design and implement databases. Students will learn how to configure database servers and, if required, learn how to develop an application cluster. They also need to create and modify databases and database components such as tables, views, stored procedures, user-defined functions, triggers, and cursors. After a database is created, students need to populate the tables with data. They will learn how to transfer data in and out of a database. Other tasks may include those that can extend the functionalities of a database such as storing, retrieving, and processing XML data. Students will learn how to query the information stored in a database with structured query language (SQL), or using other programming languages such as C++, C#, and Visual Basic to query and manage databases. Students will develop skills in database management. They will learn how to back up and restore databases, create user accounts, create roles, and grant permissions. Some of the database system courses may also require students to set up data warehouses, data marts, online analytical processing (OLAP), and cubes to support decision making.

Application Development

An application development course requires students to develop front-end projects and some application server projects on an application server. The front-end projects may include designing and implementing forms and reports. Students will learn how to access the back-end database or the midtier application server to retrieve the data to be displayed on a form and report. Students are required to develop triggers and stored procedures to automatically start some activities based on certain conditions set for forms and reports. Images are important for a form or report. Students will also learn how to display images on a form or a report. For the application server, the tasks that can be done by students are Web server implementation, programming for business logics, data transformation, Web service development and configuration, and processing XML data.

Through hands-on practice, students will have a solid understanding of the topics covered in each technology-based course and they will learn skills for solving some real-life problems. To support lab activities for the four above mentioned courses simultaneously in one semester, it requires a medium computer lab to meet the requirements of hands-on practice. Based on the above hands-on practice requirements, let us first select the hardware for the lab.

Hardware Requirements

Suppose that the budget for the online computer lab construction is limited. We cannot cover the cost for a brand new computer lab. We also cannot use the computers in the general purpose computer labs since our students need the administrator's privilege. Our solution is to check what we have in the surplus computer storage to find something useful for the lab and purchase the items that are really necessary. Suppose that currently we have desktop computers and some other equipment in the surplus storage with the following configurations.

*   Computers with a 600MHz Pentium III CPU and 256MB RAM. Each of these computers has a 20GB hard drive, a floppy-disk drive, a network interface card, and a CD ROM drive.
*   Network cables with RJ45 connection.

- A printer.
- Some CRT monitors.

Many of this equipment is valuable for our lab construction. In the following, let us take a look at where we can use this equipment.

Hardware Requirements for Networking Course

As mentioned before, the network for a networking course is unstable and cannot be shared with other courses. Since the lab network is not shared with other courses, there is no other application software installed. Therefore, the performance requirements are relatively low and we can use virtual machines to implement the network. Suppose that a class has 20 students and each student will use at least two virtual machines in the lab. We then need to create 40 virtual machines. Suppose that each virtual machine needs 256MB RAM to run a server operating system properly. The following are descriptions of three different solutions for constructing the lab.

- **Using surplus computers:** If we decide to use surplus computers, then we need to upgrade the RAM to 1GB so that each surplus computer can handle three virtual machines and one of the virtual machines can be configured as a router. In such a case, we need 20 surplus computers and 20 CRT monitors to support 40 virtual machines and 20 routers. We can either use VMware or Microsoft Virtual Server to create those virtual machines. However, on the older Pentium III computer, Microsoft Virtual Server is slower. VMware may be a better choice. Twenty surplus computers should be configured into one network which needs additional network equipment such as network cables, switches, or hubs. For this solution, the cost of hardware is minimal.

- **Using new computers:** Suppose we need better performance and the budget allows us to purchase new Pentium IV computers. Then, we may have the new computers with the configurations such as Pentium IV 3GHz CPU, 2GB RAM, 80GB hard drive, floppy drive, and CD-ROM drive. Then, we may be able to handle six virtual machines and one virtual router on each host computer. We need seven such host computers and a switch. This solution takes less space and has better performance.

- **Using server computers:** Another solution is to use a server or a set of servers installed on a rack system or in a blade enclosure. If a single server is used, we may need 12GB RAM installed on the server, which is acceptable for a server computer. To avoid bottlenecks, we can consider adding a few more network interface cards to the server, or consider using a rack system or a blade system with two or three servers. This solution takes minimal space but is the most expensive choice.

To be able to access the computer lab through the Internet, an additional surplus computer can be configured as a VPN server with two network interface cards. One of the interface cards should be configured for the local area network that connects all the computers in this lab, and the other one should be configured to have a routable IP address so that students are able to access their own host computer from home through the Internet. The VPN server should be linked to the university's network, or through DSL or cable connection. Once the virtual machines are set up, students can practice system and network configurations. Depending on the lab requirements, the network configurations can be changed for each different project.

Hardware Requirements for Network Management Course

For this type of course, the hardware requirements are similar to those of the networking course. However, you cannot share its network with the networking course since this course requires stable configurations of the network interface cards and routers. Thus, if the lab uses surplus computers, you will need another 20 computers. Or, if you decide to use new computers, you will need another seven computers. The other option is to use another server computer with 12GB RAM installed. If there is not enough room to keep all the computers to support different courses, using powerful servers to handle multiple courses is a better solution. Again, you need to configure a VPN server for remote access.

Hardware Requirements for Database System Course

In this course, database systems are constructed on a client-server structure. Again, we assume that 20 students are enrolled in the class and each student needs two computers, one used as a server and the other used as a client computer. This lab cannot be shared with those two network-related courses since it requires a stable network. The network configurations should remain the same through out the semester. Often, a database server requires more RAM to function properly. Suppose that the server operating system and database server together need 512MB RAM which is twice as much as the RAM used for each machine in a networking class. This requirement makes the surplus computer solution less attractive since we have to double the amount of RAM on each of the computers to host the virtual machines. Some of the older computers may not be able to handle more than 1GB RAM. In such a case, we may need to install 40 surplus computers and each with 512MB RAM. The installation and management of 40 computers is not an easy job.

A better solution is to use two server computers, each with 12GB RAM, multiple virtual machines, and database server software installed. On the virtual machines, students can perform tasks such as database installation and configuration. Although the network is stable, the database server is in an unstable situation. The lab also needs a VPN server for remote access.

Hardware Requirements for Application Development Course

The lab for this course is also constructed on a client-server structure. However, it cannot share the network with the database system course because it needs a stable database server. An easier solution is to install a DBMS on a server equipped with 1GB or more RAM. Create a user account for each student on the database server and make the database accessible through the Internet. In such a way, students do not have to have the administrator account for the server computer. Therefore, there is no need for virtual machines and there will be better performance. Students will use their home computers as client computers which are so configured that students can access the database server through the Internet. The Internet connection will be through a VPN server or simply a DSL router. By using students' home computers as client computers, this lab requires minimal computing resources.

At this point, we have identified all the hardware to support four courses simultaneously. This lab can be built with more than 80 surplus computers having minimal cost, or with five powerful servers which require minimal room space, or anything in between.

Software Requirements

If a computer science or information systems department is supported by an academic program from major software companies such as Microsoft MSDN Academic Alliance program or Oracle academic initiative (OAI) program, the cost of software is minimal. In the following, let us take a look at what the software requirements are for each of the courses.

Software Requirements for Networking Course

For a networking course, the main software requirement is the server operating system. On the server side, we can install Microsoft Windows Server 2003 or Red Hat Enterprise Linux 4.0 on each virtual machine. For students' hands-on practice, these server operating systems do not need to be fully configured. Be aware that Red Hat Enterprise Linux 4.0 may require more than 256MB RAM if it is fully installed. Network service packages such as DNS, DNCP, FTP, and auditing and monitoring software are included in the server operating system if it is fully installed. The client operating system can be Microsoft Windows XP Professional or Red Hat Enterprise Linux WS 4.0.

Software Requirements for Network Management Course

The software for a network management course is similar to that for the networking course. The difference is that the network interface cards, DNS, and other network devices should be already configured and should function properly.

Software Requirements for Database System Course

For a database system course, a DBMS software package should be installed on virtual machines which are run by a server operating system. A server operating system, such as Microsoft Windows Server 2003 or Red Hat Enterprise Linux 4.0, will allow multiple users to log on to the server computer at the same time. In this example, if the server operating system is Microsoft Windows Server 2003, we will install Microsoft SQL Server 2005 which can handle most of the enterprise-level hands-on practice and which is covered by the MSDNAA program. The other option is Oracle Database Server. If the server operating system is Red Hat Enterprise Linux 4.0, the choice is Oracle Database Server. If needed, the application analysis software such as online analytical processing (OLAP) and data mining should also be installed on the virtual server machines. Database design software such as Microsoft Visio for the Windows operating system or Cherokee for the Linux operating system can also be installed on the virtual server machines. For each virtual client machine, DBMS client software should be installed and configured to communicate with the virtual server machine.

Software Requirements for Application Development Course

An application development course has fewer requirements for hardware, but it has more requirements for software. The software needed by an application development course include a DBMS for storing and managing data, software for the application server such as programming language packages, Web service development tools, Web server software and so on, and software for developing front-end applications. According to the lab setup, students' home computers are used as client computers. The MSDNAA program allows students to download client-side software such as Windows XP Professional, Microsoft Access, and InfoPath and install the software on their home computers. On the server side, one can install the DBMS on one server and the application server software on a different server. If the performance is not the concern, on the server hosting the DBMS, you can also install application server software such as programming packages and office production suite packages.

Network Development

In this section, we will configure the network so that computers in the lab will be physically connected and can exchange information with each other. Depending on the how the lab is used, the network configuration can be specified as the following.

Network Development for Networking Course

As mentioned earlier, in this example, virtual machines are used for hands-on practice. Since this class requires students to construct the network, we should leave the network construction tasks to the students. One thing that needs to be done is to make sure that the host computers are connected to the VPN server so that the students can remotely access these computers. The VPN server should have at least two network interface cards installed. One of them is configured to communicate with the university's network and the other one is used to communicate with the host computers. After installing the operating systems to the virtual machines, we will leave the rest of network configuration tasks such as configuring routers and network interface cards to the students.

Network Development for Network Management Course

It is required that the network should be built and tested before the class starts. Suppose that each student has two or more virtual machines and at least two of them are virtual server machines. The lab can be networked as the following.

- A local virtual network should be constructed and each virtual server machine is assigned an IP address. For this case study, to give each student an opportunity to get hands-on practice on directory service configuration, 20 of such subvirtual networks will be built. To allow communication between the subnets, routers should be configured to route the packets from one subnet to another.

- Each virtual machine that is used as a server should have a fixed IP address 192.168.n.x where n is used for subnet identification and x is a number from 2 to 254 assigned to each individual virtual machine in the subnet.

- Configure a lab server that can communicate with each subnet. The lab server is configured as a domain controller. It is also a DHCP server and a DNS server.

- NAT should be configured so that the virtual machines can communicate with the host computer and subnets.

- If the desktop computers are used to host the virtual machines, these computers should also be networked. The desktop computers should be linked to a switch that has enough ports. If there are not enough ports to link all the host computers, more switches should be used. The network of the host computers and the VPN server should be configured so that the host computers and the internal VPN server network interface cards are in the same network.

Network Development for Database System Course

Again, assume that each student will have at least two computers; one of them is used as a server and the other one is used as a client. The server computer should have a fixed IP address and the client computers can either use a fixed IP address or a dynamic IP address automatically received from a server computer. All the computers in the lab should have the same network ID as the one that the VPN server's internal network interface card has, so that they can communicate with the VPN server for remote access.

Network Development for Application Development Course

Since we can use a single server in the lab for this course, there is not much to do for the network configuration. The network interface card for the host server should be configured so that it has the same network ID as the VPN server. This will allow students to remotely access the server from their home computers.

Final Thoughts on Network Development

As described above, all these internal networks need to access the Internet or an organization's network through a VPN server. The VPN server could have an I/O bottleneck. You may either configure multiple network interface cards, one for each internal network, or use multiple VPN servers.

After all the hardware and software are installed and the networks are configured, the next task is to create some testing accounts to test and see if the lab networks function properly. Use these testing accounts to remotely access the online computer lab and go through some of the lab activities. It is likely that you need to make some modifications based on the testing results. After all the problems are fixed, the online computer lab is ready to be used for teaching and hands-on practice.

Conclusion

This chapter has discussed issues related to the network development. It has investigated networks for different computer lab models. We have discussed the requirements for network connectivity, security, reliability, and performance; this discussion can be used as a guideline for network projects.

At the end of this chapter, a case study was given to illustrate the development of a medium computer lab. The example was based on the situation where four technology-based courses would be using the lab and each of them needed its own network. Virtual machines could be used to construct the networks in this case study. The case study provided detailed information about the configurations of hardware, software, virtual machines, and network equipment.

This chapter has accomplished the tasks related to network development and implementation. After the servers and networks have been installed, properly configured, and tested, the online computer lab is ready to support the teaching and hands-on practice required by the technology-based courses. Our next task is to discuss the issues of how to properly install and configure the student computers on the client side. We will cover these topics in the next chapter.

References

Bautts, T., Dawson, T., & Purdy, G. (2005). *Linux network administrator's guide.* Sebastopol, CA: O'Reilly Media.

Chellis, J., & Sheltz, M. (2005). *MCSA/MCSE: Windows Server 2003 network simulator.* Alameda, CA: Sybex.

Comer, D. E. (2004). *Computer networks and Internets with Internet applications.* Upper Saddle River, NJ: Prentice Hall.

Correia, E., & Watson, R. (2006). VMware as a practical learning tool. In N. Sarkar (Ed.), *Tools for teaching computer networking and hardware concepts* (pp. 338-354). Hershey, PA: Idea Group Inc.

Hassan, M., & Jain, R. (2003). *High performance TCP/IP networking: Concepts, issues, and solutions.* Upper Saddle River, NJ: Prentice Hall.

Hunt, C., & Bragg, R. (2005). *Windows Server 2003 network administration*. Sebastopol, CA: O'Reilly Media.

Schmied, G. (2004). *Integrated Cisco and UNIX network architectures*. Indianapolis, IN: Cisco Press.

Stockman, M. (2003). The remotely accessible and readily expandable networking lab: Using virtual systems technology in combination with remote control software. In G. Richards (Ed.), *Proceedings of World Conference on E-Learning in Corporate, Government, Healthcare, and Higher Education 2003* (pp. 2468-2469). Chesapeake, VA: AACE.

Teare, D., & Paquet, C. (2005). *Campus network design fundamentals*. Indianapolis, IN: Cisco Press.

Varvel, V., & Harnisch, D. (2001). The modern classroom: Using portable wireless computer networking in the classroom. In C. Crawford et al. (Eds.), *Proceedings of Society for Information Technology and Teacher Education International Conference 2001* (pp. 1363-1368). Chesapeake, VA: AACE.

Wehrle, K., Pahlke, F., Ritter, H., Muller, D., & Bechler, M. (2004). *The Linux networking architecture*. Upper Saddle River, NJ: Prentice Hall.

Chapter VII

Client Development for Online Computer Labs

Introduction

After the servers and networks have been developed, it is time to configure client computers so that they can be used to access the servers. Correctly configured client hardware and software will allow students to access the servers and complete their hands-on practice through the Internet. In this chapter, we will discuss issues related to the configuration of the client side. There are two types of client computers. Students' home computers can be used as client computers. Also, some of the computers in the online computer lab can be configured to be client computers as part of the client-server architecture. The hardware on the client side includes desktop computers, notebook computers, PDA, printers, scanners, network equipments, and audio and video devices. The client-side software includes client operating systems, remote access utility, multimedia software, programming software, office production software, security software, Web browser, and other application software.

This chapter starts with the discussion of the issues related to the configuration of client hardware and software. Next, this chapter will discuss the configuration of remote access. We will take a closer look at how to connect students' computers to the server located on campus. Then, we will discuss the issues related to the installation and configuration of audio and video devices with which students can listen and watch live demonstrations on their client computers. We will also discuss the tools for collaboration. These collaboration tools will help students with group activities.

Background

An online computer lab is constructed on client-server architecture. Networks covered in the previous chapter link servers to clients. For the development of clients, the first task is to select a proper remote access scheme. There are many ways to remotely access an online computer lab. The commonly used remote access methods are virtual private network (VPN), Terminal Services, virtual machine remote client, and database remote client. When the VPN technology is used to access a remote online computer lab, the VPN client-side software installed on students' computers needs to be properly configured (Snader, 2005). Citrix is a commercial remote access tool (Kaplan, Reeser & Wood, 2006). It is used by many companies and higher education institutions. Terminal Services is another commonly used remote access tool. In fact, Terminal Services are based on the technology developed by Citrix (Tritsch, 2005). The Terminal Services technology is included in Windows XP Professional called remote desktop connection. If Windows 2000 Professional or an older version operating system is used on the client side, one needs to install and configure the Terminal Services client software on the client's computer.

Multimedia can be used to greatly improve the quality of online teaching and learning. Instructors and students who will create and use multimedia course materials need to know how to correctly use the multimedia technology in Web-based teaching and learning (Clark & Mayer, 2003). To create multimedia course materials, instructors should know how to use multimedia authoring tools. Flash is one of the commonly used multimedia technologies (Castillo, Hancock, & Hess, 2004). The Flash technology can be used to develop multimedia e-learning materials. With Flash, one can create interactive Web pages. Castillo et al. (2004) show how to communicate with learning management systems.

Many of our students prefer learning through hands-on practice. They need to practice and try things to enhance their understanding. Tools are developed to perform and create interactive online e-learning materials (Watkins, 2005). These tools can help users to develop interactive course content with chat rooms, discussion boards and e-mail. These interactivities can greatly improve online teaching and learning. These tools can also be used to develop computer games and simulations used to assist in online teaching and learning (Aldrich, 2005). To develop games and simulations, developers need to be familiar with programming related topics such as programming preparation, coding and testing the programs for interactive games, and strategies to produce an effective learning game.

In a Web-based class, collaborative learning can be carried out to improve the quality of learning. Students often need to install and configure collaboration tools on their own computers (Rosenberg, 2006). The collaboration tools can help students and instructors to arrange and manage team-based learning through the Internet. These tools can also help the creation of an environment for team collaboration (Michaelsen, Knight, & Fink, 2002). Some advanced collaboration tools can be used for synchronous collaboration and for live training (Bonk, 2002). These tools include conferencing tools, learning management system tools, and work team tools.

For the purpose of this book, technology for online computer labs is one of the areas that we will focus on. The tasks of installing and configuring client-side technologies include identifying the technologies needed for dealing with multimedia course content, making decisions on the selection of collaboration tools, and using the technologies to remotely access online computer labs. Instructors' computers can also be considered as client computers. In addition to the above mentioned tools and technologies, more software packages are installed on instructors' computers to develop, manage, and deliver online course content (Horton & Horton, 2003).

The research on the development of client tools is quite active. Esche (2002) summarizes various remote access tools for a remote experiment in engineering education. Lister (1998) discusses the use of collaboration tools in distance learning for courses such as Hands-On Multimedia and Hands-On World Wide Web with technologies such as the Internet, Web server, remote access server, LMS, and satellite broadcasting. Lister (1998) reports that over 8,000 students enrolled in the Hands-On World Wide Web course. Multimedia tools can greatly improve the quality of e-learning. Monahan, McArdle, Bertolotto, and Mangina (2005) discuss the use of 3-D graphical user interfaces and multimedia to enhance distance learning. They describe a virtual reality environment developed to support online courses. Troupin (2000) explains how instructional design can be applied to multimedia content development. She points out that not all multimedia content is created equal. With good instructional design, the created multimedia content can enhance learning by reducing the learning stress and encouraging the learner to explore further.

Client Computer Systems

Client computers are installed on the client side of the client-server architecture; they can be in a computer lab and at students' homes, in offices, and wherever students can use them to access the servers in an online computer lab. In the computer lab, the client computers are often the same type of computers that are either new computers through purchases or old computers obtained from a surplus storage. It is relatively easy to install and configure these computers. Since these computers are often of the same type, we can use the disk imaging technique to install the operating system and application software. The trouble is that the client computers at students' homes or other places are usually different types of computers. They can be any type of computer system installed with various operating systems and different application software. It certainly complicates the client development. In the following, let us investigate some of the issues related to the development of client-side computers.

Client Computer Hardware

The requirements for the client-side computer hardware are different depending on the needs of teaching and hands-on practice. The minimal requirement for all types of client computer hardware is that these computers should be able to display the desktop image of the server computers in an online computer lab with acceptable speed. Most of the computer systems sold on the current market will satisfy this requirement. Even some of the older computers will have no problem meeting the requirement. For example, computers with the following configurations are adequate to meet the client computer hardware requirements.

- **CPU:** Pentium III with 600MHz.
- **Memory:** 256MB RAM.
- **Hard disk:** 20GB hard drive storage.
- **Portable media:** A floppy-disk drive, a flush drive, or a CD-ROM drive.
- **Network device:** Network interface card.
- **Monitor:** A CRT or LCD Monitor.

Nowadays, most of our students have computers that meet these requirements. For some courses such as network management and system administration, students only need to access the server computers in an online computer lab. The students can use even less powerful computers. In the case where a student computer is used as the client side computer of the client-server computing system architecture, the hardware requirement will depend on the requirements of the application software. Some of the client application software may require higher central processing unit (CPU) speed and others may need more memory. In general, the above computer hardware configurations are enough for client-side application software. When notebook computers are used as client computers, similar configurations can be used for these computers, except the monitors.

Client Operating Systems

The basic requirement for a client-side operating system is that it must support remote access to the servers in an online computer lab. Many of the operating systems can meet this requirement. For example, we can choose one of the Linux operating systems, Microsoft Windows XP Professional Edition, Microsoft Windows XP Media Center Edition, or Microsoft Windows Server 2003. However, students should be aware that Microsoft Windows XP Home Edition may have problems supporting the Web server's Internet information services (IIS); this may prevent students from performing application development hands-on practice. On the other hand, Windows XP Home Edition is the operating system installed on most of the computers sold on the market.

When a student purchases a new computer, it is likely that the operating system is Windows XP Home Edition. If this is the case, the student may need to replace Windows XP Home Edition with another Windows edition that meets the hands-on practice requirements. The quick and easy solution is to choose the computer model that has Windows XP Professional Edition or Windows XP Media Center Edition installed. Some vendors allow buyers to choose the operating system when they sell new computers. Make sure to inform students to choose Windows XP Professional Edition or Windows XP Media Center Edition if you use Windows operating systems.

If a student cannot choose the operating system when purchasing a new computer, or his/her computer was purchased before and Windows XP Home Edition was already installed on the computer, the student needs to replace the operating system. Suppose that the computer information systems or computer science department has a contract with MSDNAA. In this case, the student can download Windows XP Professional Edition from the university's file server. After Windows XP Professional is downloaded, there are two ways to replace Windows XP Home Edition. The student may choose to replace the existing operating system while installing the new operating system. Since there are always some compatibility problems when replacing the existing operating system, especially when many application software packages are already installed on the current computer, it is better to make a fresh installation of the new operating system and application software. A problem with this approach is that the drivers for the network card, video card, monitor, and audio devices may not match; the student may have to go to the vendors' Web sites to download compatible drivers and then manually install them. The whole process takes time and skill. If the student is not familiar with the process, he/she can get frustrated. The computer service department should prepare to offer help to students for this matter. A general guideline on replacing an operating system should be posted on the department's Web site.

After the operating system is installed, several configurations are needed to make sure that the client computer is ready for remote access. The following are some configuration related activities.

- Create a user with the username and password. The user is configured to have the administrator's privilege.
- Reconfigure the Web browser to allow the downloading and executing of ActiveX controls and Java on the client computer. If necessary, students may need to turn off the pop-up blocker since the blocker may block some Web-based applications.
- Install and configure Web server software, such as Internet Information Services (IIS).
- Configure the network card for the Internet connection.

After the operating system is properly configured, the next task is to install and configure the application software on the client computers.

Home Network

To remotely access the servers in an online computer lab, students need an Internet connection. There are several ways to connect to the Internet such as dial up, DSL, cable, satellite, and broadband wireless Internet access. Many of the lab activities require a broadband Internet connection. If possible, students should try to avoid the dialup Internet connection. The following are some guidelines for setting up the Internet connection at home.

- Search for the Internet service providers (ISPs) that cover the residential area.
- Compare the transmission speed, cost, reliability, flexibility, security, and technical support provided by those ISPs. Choose the one that is the best suitable for online hands-on practice.

- The selected ISP will send a technician to connect the home computer to the Internet through a modem and activate the Internet service.

Often, a student may have multiple computers at home and would like to access the Internet from any of those home computers. In this case, the student needs to build a small network to connect all the computers. The following are the general tasks for setting up the home network.

- Purchase an Internet router that matches the type of Internet service the student has selected.
- Physically link the home computers to the router, and then link the router to the modem used for the Internet connection.
- Configure the router's WAN option by using the network information provided by the ISP.
- Configure router's DHCP server to provide dynamic IP addresses for the home computers.
- For each of the home computers, configure the network connection to connect to the Internet router. The network can be either a wired or wireless network.
- Configure the DSL router for the Internet connection. If the VPN is used for remote access, make sure that the DSL router allows the routing of Point-to-Point Tunneling Protocol (PPTP) packets.
- Test the result of the configuration.

After successfully configuring the home network, the student should be able to access the Internet and the servers installed in the online computer lab.

Application Software

On the client computers, students can install a variety of application software. Running the application software on the client computers can gain better performance and reduce the network traffic to the online computer lab. The disadvantage is that the same application software may not always work on some client computers due to the fact that there are all kinds of computer hardware on the client side. The other disadvantage is the cost. The cost of application software will add up to a significant amount which some students may not be able to afford. Also, some application software may require a much powerful computer to run; this adds more cost for the hardware. There are three ways to solve the problem of cost.

Using Java-Based Application Software

Suppose that an application development course requires that all the application hands-on practice be done with Java based application software. Then the following software may be installed on a client computer.

- **Java 2 Platform Standard Edition (J2SE):** This package includes a collection of Java application programming interfaces (APIs) which are used to create Java programs to implement business computation logics and to create graphical user interfaces (GUIs) and network protocols. J2SE can be installed on Solaris operating systems and Windows operating systems such as Windows XP Professional and Windows 2000 Professional. For the Solaris platform, it is recommended that a computer should have 750MHz UltraSPARC III, 1GB RAM, and 810MB storage space on the hard disk. The Windows platform requires a computer to have a Pentium IV 1GHz CPU, 1GB RAM, and 880MB storage space on the hard disk. You may also need to set a path for the J2SE programming environment.

- **Java 2 Platform Enterprise Edition (J2EE):** This package is a full version Java application development package. It includes all of the classes in J2SE. In addition, it also contains classes that are more useful for developing server-side programs. The hardware requirements are similar to those of J2SE except that it needs a bigger storage space. A 10GB hard disk should be enough for full installation. Again, you may need to set a path for your programming environment.

- **Java Studio Creator:** This package is used to simplify the development of visual Web applications and the development of portlets. By using Java Studio Creator, students can easily access databases and build streamlined objects for modeling complex business domains, rules, and systems. To install Java Studio Creator, you need a powerful computer that has Linux, Solaris, or Windows XP Professional installed. The Solaris operating system requires a computer to have a 750MHz UltraSPARC III CPU, 1GB RAM, and 1.65GB space on the hard disk. The Windows platform requires a computer to have a Pentium IV 1GHz CPU, 1GB RAM, and 1GB space on the hard disk. For the Red Hat Enterprise Linux operating system, the requirements for hardware are similar to those for the Windows operating systems. To install Java Studio Creator, you are also required to have the newer version of Web server software installed on the computer.

- **Java Studio Enterprise:** This package is an integrated development environment (IDE) used to develop J2EE applications. By using this IDE, students can develop, tune, debug, and test enterprise applications, Web services, and portal components. If Java Studio runs on a Solaris system, the computer should have an UltraSPARC IIIi 1GHz CPU, 1GB RAM, and 1GB hard disk space. For Windows operating systems, it is recommended that the client computer should have a Pentium IV 1.4GHz CPU, 1GB RAM, and 1GB hard disk space. To make Java Studio Enterprise run properly, the computer should have J2SE Development Kit 5.0 Upgrade 3 or later upgrade installed.

- **StarOffice:** StarOffice is a product that is similar to Microsoft Office. It includes full-powered word processing, spreadsheet, presentation, drawing, and database capabilities. If you use Solaris as the operating system, you will need 128MB RAM and 300MB hard disk space to run StarOffice. It is recommended that an X -window have 800x600 or higher resolution with 256 colors. For the Linux operating system with Linux Kernel version 2.2.13 or higher, StarOffice will require any Intel Pentium compatible processor, 128MB RAM, and 300MB hard disk space. The Linux operating system should include the desktop GUI tool GNOME 2.0 or higher.

The above are application development tools from Sun. These application development tools, except StarOffice, can be downloaded from the Sun Web site for free. For educational institutions, StarOffice is also free. These application development tools are compatible with the Linux operating system. If the client computer has the Linux operating system installed, these tools are the right choice since most of the Microsoft application development tools are not compatible with the Linux operating system.

Using Microsoft Application Software

If the client computer has one of the Windows operating systems installed, there is plenty of Microsoft application software. If a computer information system or a computer science department is under the MSDNAA contract, the software will be free to students and faculty members in the same department. Otherwise, the cost of the software can be expensive for some students. With the MSDNAA contract, the following application software can be downloaded from a file server on campus to the students' home computers.

- **Microsoft Visual Studio .NET:** This package is an integrated development environment. By using Microsoft Visual Studio .NET, students can create programs in various programming languages, construct Web sites, develop Web applications, and create Web services that run on the Internet. Multiple programming languages such as Visual Basic .NET, Visual C#, Visual C++, and Visual J++ can share the same project development environment. Visual

Studio .NET is a large software package. It requires the client computer to be equipped with a 1GHz Pentium III or higher CPU, 256MB RAM, 1GB hard disk space, and 2GB space for installation. The surplus computers mentioned in the earlier examples satisfy the minimum requirement. It also requires that the IIS be installed and configured for this package. If students want to install the MSDN documents, it will take another 3GB hard disk space.

- **Microsoft Project Professional:** This package is used for project management. Students can use this software for courses such as software management where they will use the software to develop plans, allocate resources, track a project's progress, manage the project's budget, and analyze workloads for the project. The requirements for running Microsoft Project Professional are that the client computer should have a Pentium III CPU, 128MB RAM, and 130MB hard disk space.

- **Microsoft Visio:** This is a software package that is used to model business processes. Visio 2003 can be used to create data models in a database design and development course. It can also produce diagrams for other courses too, for example, creation of network topologies for a networking class or flow charts for a programming course. The requirements for the client computer are similar to those of Microsoft Project Professional except that Visio requires 220MB hard disk space which is a little more than the capacity of the hard disk required for Microsoft Project Professional.

- **Microsoft OneNote:** When students participate in a live discussion group or a live tutoring session, OneNote can be used to organize, capture, deploy, and share hand-written notes in presentations. The requirements for running Microsoft OneNote are also similar to those for Microsoft Project, that is, a client computer with a Pentium III CPU, 128MB RAM, and 130MB hard disk space.

- **Microsoft InfoPath:** InfoPath is used to develop applications such as forms and reports in XML. It can be used to view XML documents. It can consume the XML data provided by XML Web services through MSXML and SOAP Toolkit. Again, it requires the client computer to have a Pentium III CPU, 128MB RAM, and 100MB hard disk space.

- **Microsoft Internet Information Services (IIS):** This application software is a Web server that can be used to manage and scale Web application infrastructure. Usually, it is bundled with the operating system software such as Microsoft Widows XP Professional Edition. When Microsoft Visual Studio .NET is installed, you need to properly configure IIS for ASP.NET and Web services components. You can use the Add and Remove Programs wizard to install IIS. In general, the requirement of computer system resources for installing IIS is minimal.

Using Tools Included in DBMS Package

Many front-end applications can also be done with tools included in a database management system (DBMS) package. Microsoft Access is a personal database package. It can also be used as an application development tool. The following are some of the things Microsoft Access can accomplish.

- Microsoft Access can be used to connect to a SQL Server database at the back-end.
- After the connection to the back-end database is established, Access can retrieve data from the SQL Server database and use the data to support forms, reports, and other database applications.
- Access supports the Visual Basic programming language which can be used for implementing more complicated business logics.

To install and run Access, a computer needs to have a Pentium III CPU, 128MB RAM, and 180MB hard disk space.

Oracle also provides application development tools. To develop applications for Oracle databases or for Web-based applications, students need to install the Oracle Developer Suite 10g package on their home computers. Oracle Developer Suite 10g includes the following tools.

- **Oracle JDeveloper:** This tool is used to model, develop, debug, optimize, and deploy J2EE applications and Web services.
- **Oracle Designer:** This tool is used to design enterprise-level databases and applications.
- **Oracle Forms Developer:** This tool is used to build Internet enabled forms for an application project.
- **Oracle Software Configuration Manager:** This tool is used to manage data and files throughout a development cycle.
- **Oracle Reports Developer:** This tool is used to build Internet enabled reports for an application project.
- **Oracle Discoverer:** This is a tool that can be used to perform ad-hoc queries and analyze report results.
- **Oracle Warehouse Builder:** This tool is used to build data warehouses to support business intelligence projects.
- **Oracle Business Intelligence Beans:** This package includes a set of predeveloped Java beans for the fast development of business intelligence applications.

The Oracle Developer Suite 10g package requires that a computer at least have 256MB RAM and 2GB hard disk space.

There are many application development tools included in Microsoft SQL Server as well. The following are some of the commonly used tools:

- **Common Language Runtime Integration:** This tool allows students to develop database objects by using any Microsoft .NET language.

- **XML integration:** Several XML development tools such as XML Path and XQuery are included in SQL Server 2005 to create new XML data types, and store and retrieve XML fragments or documents in SQL Server databases.

- **Analysis Services:** This tool can be used to develop online analytical processing (OLAP) and data warehousing projects for decision support.

- **SQL Server Integration Services (SSIS):** This tool is used for extracting, transforming, and loading packages for data warehousing.

- **Reporting Services:** This tool is used for building, managing, and deploying enterprise reports.

- **Data Mining:** This tool can be used to identify the trends in business processes and the patterns in customer behavior.

To run SQL Server 2005 with the above components, it is required that the client computer should have a 600MHz Pentium III-compatible or faster CPU, 512MB RAM, and 350MB hard disk space.

To make the installation and configuration of application software go smoothly, students should follow the above recommendation. There are other types of application software such as multimedia software, remote access software, and collaboration software. Those types of the application software will be covered in the next few paragraphs.

Remote Access Configuration

In Chapter IV, we discussed several ways to remotely access the servers in an online computer lab. In this chapter, let us take a look at the client-side configuration for these remote access technologies.

Virtual Private Network Client Configuration

A commonly used method to remotely access the servers installed in an online computer lab is through the virtual private network (VPN) technology. Students can download the VPN software from the vendor's Web site and configure the VPN client on their home computers. Some of the operating systems include the VPN components. If so, after the client operating system has been installed, a student can configure the VPN to remotely access the computer lab servers.

VPN on Red Hat Linux Operating System

For a client computer installed with a Red Hat Linux operating system, the tasks in the configuration of the client VPN are shown below.

- Install the dynamic kernel module support package and point-to-point protocol (PPP) kernel module. PPP is an Internet protocol commonly used to establish a direct connection between two nodes.
- Install the GUI tool PPTP Client.
- Install php-pcntl which is used to assist the GUI tool PPTP by adding process control functions to PHP and php-gtk-pcntl which can be used to simplify the writing of client-side cross-platform GUI applications.
- Install the pptpconfig package which can be used to install Point-to-Point Tunneling Protocol (PPTP). PPTP is an Internet protocol to allow remote users to access corporate networks securely across the Internet.
- Run the pptpconfig package to start the VPN configuration GUI.
- Configure the VPN connection by specifying the name of the connection, the name or IP address of the VPN server.
- Connect to the VPN server with the user name, password, and domain name.

For Linux operating systems from other distributions, you may find a similar process from the distributions' Web site. Or, you can download the VPN client software and configure the VPN client accordingly.

VPN on Windows Operating System

For Windows XP Professional Edition and Windows Server 2003, the VPN utility is included in the operating system software. The following are some tasks in VPN configuration for a client computer.

- Make sure that the Internet connection is working properly and point-to-point tunneling protocol (PPTP) forwarding is enabled in the DSL router.
- Start the network connections utility which can be found in the control panel.
- Specify the connection type to be VPN connection.
- Specify the connection name.
- Choose the VPN server created in the online computer lab. You can either use the host name or use the server's IP address.
- Students need the user name and password created on the VPN server to log on to the lab VPN server. When entering the user name and password, a student may also need to enter the domain name if his/her account is managed by a domain controller.

After logging on to the lab VPN server, a student can then access each individual computer in the lab. We will discuss this in the next paragraph.

Terminal Services Client Configuration

If the client computers have Windows XP Professional installed and the servers have Window Server 2003 installed, students may be able to access the servers through Terminal Services. Since students need the administrator's privilege to the Terminal Services enabled server and there is a limit on how many administrators can log on to the server through Terminal Services, this way to remotely access the server is only suitable for the networking or system administration course where each student is the administrator for his/her own server. The tasks in Terminal Services Client configuration are given below.

- If an older version of the Windows operating systems, such as Windows 2000, is running on the client computer, you need to download the Remote Desktop Connection software from the Microsoft Web site. If Windows XP Professional is installed, the configuration can be done through the Remote Desktop Connection utility.

- Configure the servers in the online computer lab to allow the client to remotely access it through Terminal Services.
- Start the Remote Desktop Connection utility on the client computer.
- Specify the server name for Remote Desktop Connection.
- Use the user name and password of the server computer to log on to the server.

After logging on to the server, students can perform the hands-on practice as if they are sitting in front of the server. This method requires the server computer to have a routable IP address. If there are not enough routable IP addresses, students may have to access these servers through VPN.

Virtual Machine Client Configuration

If a server in an online computer lab has virtual machines installed, students can remotely access the virtual machines through the virtual machine client software. Two commonly used virtual machines software packages are Microsoft Virtual Server and VMware from EMC. The discussion of the configuration of virtual machine clients for both of these two virtual machine packages are given in the following.

Microsoft Virtual Server

If Microsoft Virtual Server is installed in the online computer lab, students need to install the virtual machine remote control (VMRC) client software provided by the virtual server to remotely access the virtual machines. The VMRC client uses the VMRC protocol to access the virtual server through the Internet. The VMRC protocol is designed for viewing and controlling virtual machines across the Internet. The following briefly describes the tasks for the configuration of the VMRC client.

- Install Microsoft Virtual Server on the host server which is running the Microsoft Windows Server 2003 operating system.
- Start the virtual server administration Web page.
- Configure the virtual server properties to enable the VMRC client.
- Compress the VMRC client folder and copy it to a portable disk or make it available for downloading.
- Install and run the VMRC client software on client computers.
- Specify the virtual server's host server's name or IP address.

- Specify the port number as 5900.
- Specify the default screen resolution.
- Log on to the host server with the user name, password, and domain name for an account created on the host server. The account for the host server does not have to have the administrator's privilege.
- Log on to the virtual machine with the virtual machine's administrator account. With the administrator's privilege, students will be able to carry out the server-side practice.

Again, the host server needs to have a routable IP address. The virtual machines do not need as many routable IP addresses as those required by Terminal Services since the host server can have multiple virtual machines installed on it. If there are not enough routable IP addresses for multiple host servers, remote access to these host servers needs to go through VPN.

VMware

VMware is another virtual machine technology. Accessing remotely to a VWware virtual machine can be done through the Remote Console software. The Remote Console software can be installed on the client computers with the Windows XP, Windows 2000, and Linux operating systems installed. The following describes the configurations of Remote Console.

- Install the Remote Console software on client computers.
- Start VWware Remote Console.
- Specify the name of the host server on which the VWware server is installed.
- Specify the log on user name, password, and domain name for the host server.
- After the host server is connected, log on to the virtual machine with the user name and password of an administrator's account.

Like Terminal Services, VWware has two more remote access tools that are worth mentioning. These two remote access tools are VMware ACE and VMware virtual desktop infrastructure. Through the two tools, students can also remotely access a host server. A brief description of these two remote access tools is given below.

- **VMware ACE:** Since VMware ACE can preconfigure a desktop with different levels of protection for remote access, the desktop is a trusted desktop with better security measures. The remote access interacts with the desktop; this prevents users' direct access to the online computer lab's network. Therefore, VMware ACE can prevent the viruses or Spyware from entering the online computer lab.

- **Virtual desktop infrastructure:** Another way to remotely access a server in the online computer lab is through the virtual desktop infrastructure which allows the client computers to host the lab server's desktop environment. VMware Virtual Desktop Infrastructure provides PC environments which are familiar to the students. The virtual desktop infrastructure solution can significantly reduce learning anxiety.

Remote Access to Application Software

If the hands-on practice only requires students to access certain application software such as a database management system (DBMS) or a learning management system (LMS) and if the application software provides the remote access components, students can remotely access the application software without having a user account in the host computer. The following are some commonly used application software packages that allow remote access.

Often the major DBMS software, such as Oracle Database and Microsoft SQL Server, allows direct remote access for database development and management. The following is the configuration information about the remote access client software for some DBMSs.

Oracle Database

To remotely access an Oracle database from students' home computers, students need to configure the connection through the Oracle Net Configuration tool. The configuration may include the following major tasks:

- Install the Oracle Client Suite package.
- Specify the Oracle database service name.
- Specify transmission control protocol (TCP) as the network protocol.
- Specify the name of the host server where the Oracle database is installed.
- Specify the user name and password for an Oracle database account.
- Specify the network service name.

After the remote access is properly configured, students are able to log on to the Oracle database server through the Internet and develop database objects.

Microsoft SQL Server

Similarly, students can access a database created on Microsoft SQL Server. The following are the tasks to configure the connection for Microsoft SQL Server 2005.

- Configure the firewall to unblock the communication with SQL Server. The default TCP port that SQL Server uses is 1433.
- Verify that SQL Server is running.
- Start the SQL Computer Manager tool and enable TCP and Named Pipes (NP) for the server and client.
- Start the SQL Browser service which is used to identify the ports that are listened to by named instances.
- Start SQL Server Management Studio on the client computer to connect to the server that hosts SQL Server 2005.

After SQL Server is connected, students can perform hands-on practice such as retrieving information from databases, managing the databases, analyzing the data stored in the databases, and creating database objects such as tables and stored procedures.

Client Multimedia Devices and Software

To participate in a live tutorial session or to review a multimedia based lab manual, students need to have audio and video devices and multimedia software installed on their home computers. If it is possible, the client side should have the following multimedia hardware and software.

Audio and Video Devices

To handle multimedia course materials, the client computers must have audio and video devices installed. These devices include a sound card, video card, speaker, microphone, camera, and monitor. The following are some configuration issues with the audio and video devices.

Sound Card

When a student purchases a personal computer, it is most likely that a sound card is already installed and configured for use. Problems may arise when the student installs a new operating system on the computer with the existing sound card. The following are some guidelines for checking the sound card.

- Check the model, the type, and the chip of the sound card installed on the client computer.
- From the operating system vendor's Web site, check if the sound card is supported by the new operating system. If not, find out which driver can be used to support the sound card on the client computer.
- Download and install the driver on the client computer.
- Make sure that the sound card driver is working properly by using a device management tool.
- Make sure that the sound card is not set at the mute mode.
- Test the sound card with some multimedia course materials.

Video Card

Most of the recently purchased desktop or notebook computers already have a good quality video card. Again, students may have problems when they try to install a new operating system on a client computer with the existing video card. The following are some tips for solving the problems.

- On the client computer, check the model and type of the video card.
- Verify if the video card is supported by the new operating system. If not, find and download the driver.
- Install the driver on the client computer and make sure that the driver is working properly.
- Properly set the screen resolution.
- Test the video card to see if it meets the requirements.

Microphone and Speaker

Usually, the speaker on a client computer is good enough to handle the multimedia course material. To record voice, a student needs a microphone set up properly on the client computer. The following are some general considerations on setting up a microphone.

- After the new multimedia software is installed, start the sound and audio device utility configuration tool.
- Change the volume for the audio recording device.
- If it is necessary, change the voice sample rate.
- If it is necessary, turn hardware acceleration on or off for better sound quality. Hardware acceleration is used for hardware to perform some functions that may gain better performance than using software.
- Test the microphone to see if the result of configuration is satisfactory.

Web Camera

A Web camera can be used for live conferencing and classroom observation. The installation and configuration of a Web camera is relatively simple. The tasks for settting up a Web camera are given below.

- Choose a Web camera that has 640x480 or higher resolution, 24-bit color, 30fps frame rate, automatically adjusts contrast and brightness, and has camera management software.
- Check the computer hardware to see if it meets the requirements for installing a Web camera. Usually, a Web camera needs to have a computer with a 200MHz or higher Pentium III CPU, USB port, 24-bits color, 32MB RAM, 150MB hard disk space, and a high-speed Internet connection. A less powerful computer may cause washed out images.
- Install Web camera software.
- When prompted, connect the Web camera to the client computer.
- Adjust the focus, contrast, and brightness.
- Use the management software to set up live conferencing. Students need to enter the information of the computer name, user name, and password for the host computer.

Streaming Media

Although video and audio are great for live conferencing, they take a lot of resources and bandwidth to work properly. Streaming video and audio are sequences of compressed signals transmitted over the Internet. The advantage of streaming video and audio is that the media is sent in a continuous stream. As soon as the media come to a viewer's computer, they will be played so that the viewer does not have to wait until the entire media file is downloaded.

There are two ways to stream audio and video; each has its own advantages. The description of these two streaming methods is given below.

- **Streaming server:** This method uses a streaming server. The connection between the client computer and the streaming server is maintained during a streaming process. In such a way, the media data can be viewed and listened to in real time. Therefore, this method of streaming is called true streaming. One of the true streaming advantages is that it allows students to fast forward and rewind the audio and video content. This advantage of true streaming is especially good for online lectures, online lab demonstrations, and live tutorials. To maintain the connection between the streaming server and the client computer, it has to use a specific protocol such as Real Time Streaming Protocol. Since true streaming automatically matches the bandwidth of the connection between the client computer and the streaming server, the quality of true streaming varies widely with connection speed. It is necessary for students to have a high speed Internet connection to run the true streaming.

- **HTTP streaming:** HTTP streaming is good for smaller media files. It requires the media content to be downloaded before the viewer can see the media content. With this method, the media files are downloaded to the client computer. During the downloading process, students will be able to view the part of the media content that has been downloaded. The media files are deployed by standard HTTP servers. Therefore, there is no need to develop a specific protocol to transfer the media file over the Internet. Some of the media players are built to be able to handle the media file downloading process.

The requirements for developing streaming media are listed below.

- To keep good image quality, it is recommended that you use a high-speed Internet connection such as DSL, cable, or T1.

- It is recommended that you use a static routable IP address for the host computer.

- The requirement for the client computer is that it should have a 200MHz or higher Pentium III CPU, 32MB or more RAM, a full duplex sound card and speaker, and a video card that can handle 65,000 or more colors.

- You need a Web browser that supports audio and video streaming.

- Make sure that the firewall does not block the port used for media streaming.
- A media player should be installed on the client computer.

Once the requirements are met, you can perform the following tasks to create streaming media.

- Create a digitized audio and video data file, such as a file that contains the video and audio data captured by a digital camcorder.
- Convert the digitized audio and video data file into a streaming data file.
- For true streaming, transfer the streaming data file to a streaming server. If HTTP streaming will be carried out, transfer the streaming media file to a HTTP server or a FTP server.
- Create an instruction Web page that has a link to the streaming data file on the streaming server, HTTP server, or FTP server.
- Upload the instruction Web page to the course Web site.
- Let the students go to the course Web site, browse to the instruction Web page, and run the streaming file.

The selection of a streaming method will depend on the computing environment. If there is no streaming server set up for the class or many students are still using 56K dialup modems, the easy way for streaming is to use HTTP streaming.

Multimedia Client Software

A multimedia-based lecture requires multimedia software be installed on client computers to handle the video and audio content. The client-side multimedia software can be used to perform the following tasks.

- Play multimedia course content.
- Create multimedia course materials.
- Edit multimedia files.
- Convert one type of multimedia data to another type of multimedia data.

In the following, we will discuss some installation and configuration issues about the software that can accomplish the above tasks.

Media Player

Some of the Web browsers have the capability of handling simple media files. To handle more sophisticated media files, media players should be installed on client computers. The following are some configuration tips.

- Check the client computer to see if it satisfies the requirements. A computer with a 200MHz CPU, 64MB RAM, 200MB hard disk space, audio and video devices, an Internet connection and Web browser should meet the minimal requirements. For better media quality, a much faster computer with much more RAM is needed.

- Make sure that the drivers for the video and sound cards are compatible with the media player to be installed.

- Specify the Web browser and media file format to be used on the client computer.

- Specify the hard disk space that can be used to store the media files.

- Specify the security policies for the media player.

- Download and install the media player on the client computer.

- Add media files to the media library.

- Test the media files and troubleshoot any problems if any.

There are many types of media players; their requirements for hardware and software are different. For example, some of the media players only require a computer to have 16MB RAM. For more details, you need to search for the information provided by these media players' Web sites.

Multimedia Authoring

There are different types of multimedia authoring software packages on the market. Some of them are stand-alone ones, some are part of the Web development tools, and others are provided by the learning management system (LMS) vendors to help instructors and students create their multimedia files. To properly use multimedia authoring software, you may need to specify the following information:

- The type of browser on a client computer to be used for running the multimedia file.

- The level of security that will allow the media files to be played on the client computer.

- The type of media player required for the client computer.
- The type of media files needed to match the media player on the client computer.
- The way in which the media files can be deployed so that students can access these files.

If a university has already installed a specific LMS and the LMS provides multimedia authoring software, it will be convenient to use these tools for developing and editing the multimedia files. Otherwise, commonly available multimedia authoring tools such as Macromedia Dreamweaver can also to get the job done. The following are some tasks for configuring multimedia authoring tools:

- Make sure to close the currently active files. You may also need to stop the antivirus software.
- Make sure that the drivers for sound and video cards are updated so that they match the audio and video devices.
- Make sure that the host computer meets the hardware requirements. Usually, a multimedia authoring tool may require a 233MHz or higher CPU, 128MB or more RAM, 200MB or more hard disk space, and a monitor with 1024x768 resolution.
- Install the multimedia authoring tool and specify the type of installation and version of the package to be installed.
- Test the newly installed multimedia authoring tool.

For technology-based courses, it is most likely that instructors will use a multimedia authoring tool to create multimedia course materials. For example, an instructor can use the free download such as Breeze presenter to create PowerPoint presentations with audio explanations. Students need multimedia authoring tools too if a course such as multimedia computing requires students to create multimedia based Web pages. It is better to have all the client computers to use the same type of multimedia authoring tool. Otherwise, each student has to figure out by himself/herself how to install, configure, and manage the multimedia software.

Client Collaboration Tools

One of the difficulties for online teaching and learning is carrying out group activities. Various collaboration tools have been developed to improve the quality of teaching.

As the advance of these collaboration tools, nowadays, collaborative learning can be done through Web conferencing. At a Web conference, the attendees can talk to one another through microphones and see one another through Web cameras. They can also communicate through text messaging or e-mail. With Web conferencing or with other collaboration tools, students can exchange ideas with their classmates in a chat room, join a discussion group, or participate in live tutoring sessions. In the following, let us discuss some of the collaboration tools that can be used to teach technology-based courses online.

Web Conferencing Software

Web conferencing is a useful tool to assist collaboration among students and instructors. Web conferencing software may include the following components:

- **Interactive tools:** In Web conferencing software, the interactive tools may include virtual whiteboards, Web slides, snapshots, annotations, chat, and question manager.

- **Document viewer:** Through the document viewer, students can view the multimedia files. This is a great tool for instructors and students to present their multimedia course materials.

- **Desktop sharing:** Desktop sharing allows the viewers to view the desktop on one of the student's computer. It is a great tool for an instructor to help a student to debug a problem raised in the hands-on practice.

- **Attendee manager:** This component can be used to manage attendees such as approval of an attendee to participate in the conference, or allow an attendee to control the local desktop screen.

- **Recording:** The recording component can be used to record a live conference and save it for later replay.

- **Custom invitations:** Send e-mail or phone messages to invite students to join a meeting.

- **Meeting registration:** This component can be used to register students for a meeting.

- **Meeting scheduler:** This component can be used to schedule a meeting and to inform the participants about the meeting schedule.

- **Audio conferencing call:** This component is used to control who gets to talk during a meeting.

Web conferencing software is usually installed on a media server. Through the Internet, students can access the media server by using a user name and password created on the Web conferencing software. To assist customers in creating multimedia content for a Web meeting, major Web conferencing vendors provide add-in software, such as Microsoft Live Meeting Add-in Pack and Macromedia Presenter. The add-in software can be added to PowerPoint to produce multimedia presentations. During a meeting, presentations can be uploaded to a Web conference for demonstration. The tasks for configuring PowerPoint add-ins are given in the following:

- Make sure that the audio card, video card, and microphone on the client computer are working properly.
- Download the PowerPoint add-in file and start the installation process.
- Specify the folder to store the add-in file and complete the installation.
- Start Microsoft Office PowerPoint; the add-in menu should be found on the menu bar.
- Test the add-in by creating a multimedia PowerPoint presentation.

Web conferencing is a great tool to overcome some of the disadvantages of Web-based teaching. Besides Web conferencing, there are some other tools that are helpful for online collaborative learning.

Intelligent Tutoring Systems

To improve the understanding of course content, an intelligent tutoring system will test students with some questions about the course content. After a student submits his/her answers to the questions, the intelligent tutoring system will analyze the answers to find in which knowledge area(s) the student is lacking based on the mistakes. After figuring out what the student needs to know, the intelligent tutoring system will select a set of topics for the student to review. Once the student completes the review, he/she will be tested again.

With data collected from students, an intelligent tutoring system can develop a learning model for each student. Based on the learning style(s) of each individual student, an intelligent tutoring system can create an instruction plan just for that student. The use of intelligent tutoring systems is still in the early stage. There are still a lot of areas that need to be improved. Many of the topics in this area are still under research. Some universities and research institutions have implemented intelligent tutoring systems. There are few such systems available as commercial products. Some of the recent research projects by Tutor Research Group (2006) have been listed on their Web site.

Blog

Blog can be another useful tool for online collaborative learning. When a student posts his/her ideas about a project on a blog, other students are invited to contribute their comments. Instructors can also post discussion topics on a blog and invite students to give their opinions. It is relatively easy to set up a blog and it is much easier to maintain a blog than a regular Web site. The following are some tasks for developing a blog on a host computer.

- Choose a Web site that can host blogs. The Web site can be owned by a university or it can be a public Web site owned by Yahoo or MSN.
- Sign up to the Web site that hosts the blog that is being created.
- Specify the name of the blog and create the URL link so that other students can visit the blog.
- Create a personal profile about the owner of the blog.
- Create the blog environment by adding images and sound effects to the blog.
- Upload the precreated files to the blog content. Later, students can update the content from anywhere.
- Enable e-mail publishing for instant uploading from devices such as a camera phone.
- Test the blog to make sure that it is working properly.
- Once the blog is created, make sure to invite other students to visit the blog and update the blog whenever new information is available for posting.

Many of the collaboration tools are included in LMS packages. Collaboration tools in LMSs provide an environment that allows teaching and learning through online communication, which is a key component in the learning process.

Conclusion

In this chapter, the issues related to hardware and software development, installation, and configuration on the client side have been discussed. First, we discussed the requirements for the client-side computer systems. Then, we discussed the issues related to client operating systems. The chapter has briefly described the operating systems that are suitable for client computers. It has discussed some tasks related to configuration of client operating systems. After the discussion of client operating systems, we focused on the issues about linking students' home computers to the Internet. Next, we dealt with the configuration of client-side application software. The application software discussed in this chapter includes Java based application software such as J2SE, J2EE, Java Studio Creator, Java Studio Enterprise, and StarOffice. We also talked about the application software from Microsoft such as Microsoft Visio, Microsoft Project, Microsoft Visual Studio .NET, Microsoft Info-Path, and Microsoft OneNote. The configuration of the client-side DBMS software such as Microsoft Access, SQL Server client, and Oracle Developer Suite 10g have also been discussed in this chapter.

Several remote access tools such as VPN, Terminal Services, and virtual machine client are discussed in this chapter. The requirements and configuration for setting up these remote access tools have also been given in this chapter. Some of the DBMS software such as Oracle Database and SQL Server allow users to log on from a client computer directly to a database without a user account for the host operating system. The configuration for this type of software has also been discussed.

The next topic covered in this chapter was on the requirements and configuration of multimedia software. The issues related to the configuration of audio and video devices such as a sound card, video card, microphone, and Web camera have been presented here. We have also discussed the client-side multimedia software such as media players and media authoring tools. This chapter has provided some information on how to create streaming files as well. The reference on the role of instructional design in multimedia development is also included in this chapter.

The last topic covered in this chapter is about the collaboration tools. The information on the configuration of collaboration tools such as Web conferencing and a blog is given. We have also discussed how students can participate in a live Web conference and how they can set up a personal blog.

After the client computers have the necessary hardware and software properly configured, our next task in online computer lab development is to deal with security issues, covered in the next chapter.

References

Aldrich, C. (2005). *Learning by doing: A comprehensive guide to simulations, computer games, and pedagogy in e-learning and other educational experiences.* San Francisco: John Wiley & Sons.

Bonk, C. J. (2002). Collaborative tools for e-learning. *Chief Learning Officer.* Retrieved April 6, 2007, from http://www.clomedia.com/content/templates/clo_feature.asp?articleid=41&zoneid=30

Castillo, S., Hancock, S., & Hess, G. (2004). *Using Flash MX to create e-learning.* Vancouver, WA: Rapid Intake Press.

Clark, R. C., & Mayer, R. E. (2003). *E-learning and the science of instruction: Proven guidelines for consumers and designers of multimedia learning.* San Francisco: Pfeiffer.

Esche, S. K. (2002). Remote experimentation: One building block in online engineering education. In *Proceedings of the 2002 ASEE/SEFI/TUB Colloquium.* Retrieved April 6, 2007, from http://www.asee.org/conferences/international/papers/upload/Remote-Experimentation-One-Building-Block-in-Online-Engineering-Education.pdf

Horton, W., & Horton, K. (2003). *E-learning tools and technologies: A consumer's guide for trainers, teachers, educators, and instructional designers.* Indianapolis, IN: Wiley.

Kaplan, S., Reeser, T., & Wood, A. (2006). *Citrix Access Suite 4 for Windows Server 2003: The official guide* (3rd ed.). Emeryville, CA: McGraw-Hill Osborne Media.

Lister, B. C. (1998). Interactive distance learning: The virtual studio classroom. In *Proceedings of the Third International IEEE Conference on Multimedia, Engineering and Education*, Hong Kong, China.

Michaelsen, L. K., Knight, A. B., & Fink, L. D. (Eds.). (2002). *Team-based learning: A transformative use of small groups.* Westport, CT: Praeger.

Monahan, T., McArdle, G., Bertolotto, M., & Mangina, E. (2005). 3D user interfaces and multimedia in e-learning. In *Proceedings of the World Conference on Educational Multimedia, Hypermedia & Telecommunications, ED-MEDIA.* Chesapeake, VA: AACE.

Rosenberg, M. J. (2006). *Beyond e-learning: Approaches and technologies to enhance organizational knowledge, learning, and performance.* San Francisco: Pfeiffer.

Snader, J. C. (2005). *VPNs Illustrated: Tunnels, VPNs, and IPsec.* Boston: Addison-Wesley Professional.

Tritsch, B. (2005). *Microsoft Windows Server 2003 Terminal Services.* Redmond, WA: Microsoft Press.

Troupin, P. (2000). The role of instructional design in multimedia development. *ASTD's Source for E-learning.* Retrieved April 6, 2007, from http://www.learningcircuits.org/2000/feb2000/Troupin.htm

Tutor Research Group. (2006). Tutor research group. Retrieved April 6, 2007, from http://web.cs.wpi.edu/Research/trg/

Watkins, R. (2005). *75 e-learning activities: Making online learning interactive.* San Francisco: John Wiley & Sons.

Chapter VIII

Online Computer Lab Security

Introduction

Before an online computer lab is ready for students to perform hands-on practice, security measures need to be enforced to protect the computer lab and even more importantly to protect the university's network. Since the computer lab is online, it is exposed to hundreds of malicious viruses. Once a computer is infected by viruses, the viruses will damage files, multiply themselves to occupy the memory and hard disk space, or take away the central processing unit (CPU) process power and make the computer run very slowly.

When uploading homework assignments, students may spread certain kinds of viruses through e-mail. During hands-on practice, students may need to download some files from other Web sites; it is possible that they may download some Trojan viruses. When activated, the Trojan virus may delete files on a hard drive or send some sensitive information to others.

If an online computer lab is not properly protected, hackers may be able to log on to its computer systems to change the lab settings, delete system files, or simply use the lab resources for their own purposes. It will be even more serious if hackers access the university's network through the online computer lab. The data owned by some offices such as accounting are much more sensitive than the data used in a computer lab. Sometimes, if the network is not properly configured, even some students may be able to access the sensitive data transferred over the university's network.

In the online computer lab, damages caused by security problems often lead to the shutdown of the entire lab. It will take some time to recover the lab. If the university's network is harmed by a security problem, not only can it cause a huge loss, but it may also trigger the university to shut down the online computer lab permanently. Therefore, the online computer lab needs to be well protected to prevent serious security problems. On the other hand, rigorous security measures may cause some conflicts since students need the administrator's privilege to perform certain hands-on practice. In this chapter, we will deal with some of the security issues and come up with some solutions to the problems.

To make online computer labs secure, the first job is to figure out what to protect. Network managers need to investigate the potential vulnerability in an online computer lab. The results of the investigation should be used as a guideline for security policies developed by a team of network managers, instructors, and university administrators. The security policies should cover the issues related to the availability, reliability, integrity, and confidentiality of the computer systems and networks used by an online computer lab. We will have an overview about security technologies and security measures imposed by these technologies. The last topic to be covered in this chapter will be security management. We will discuss management planning and the tools used to implement the management plan.

Background

Security issues in e-learning are one of the greatest concerns among many higher education institutions (Weippl, 2005). It takes joint effort from university administrators, lab managers, faculty members, and students to keep online computer labs secure. All the people involved in the development of online computer labs should understand how security problems threaten e-learning projects and how to protect e-learning projects from these threats (El-Khatib, Korba, Xu, & Yee, 2003). To successfully run an online computer lab, we must consider the issues such as privacy and security. Policy makers need to fully understand the privacy requirements and security assessment. They also need to understand security technologies for enforcing security measures.

To keep computer systems and networks secure, the first task is to identify security vulnerabilities (McNab, 2004). Based on the findings, strategies can then be defined to eliminate potential vulnerabilities. We will discuss how security standards can be used to support teaching and hands-on practice. It is necessary to develop a well-defined security policy for an organization (Peltier, 2004). The security policy should be written to meet the goals identified by the security strategy (Barman, 2001). After a security policy is developed, it should be carefully carried out and tested (Kaeo, 2003).

There are many security technologies that can be used to define, configure, and enforce security measures. One of the network manager's jobs is to investigate the relationships among these network security technologies and combine these technologies to form an integrated network security structure (Fung, 2004). For each different operating system, the security design and the implementation of security strategies may vary. Windows Server 2003 and Linux are commonly used operating systems. Readers can find more information about security technology for Windows Server 2003 in the book by Clercq (2004). To find detailed information about the Linux security technology tools, readers can refer to the book by Smith (2005).

Security management is another important subject for keeping an online computer lab safe. Security management deals with various aspects such as business and legal issues of a security management process, implementing security policies, and troubleshooting security problems, protecting operating systems, networks, and applications from virus attacks, educating lab users, and fixing damage caused by security problems (Ortmeier, 2004). For the Windows Server 2003 security management technologies tools, readers can find more information in the book by Kaufmann and English (2004). If an online computer lab is built on a Linux platform, the book by Turnbull (2005) provides detailed information about the security management tools.

Yang and Dark (2005) report the implementation of an information security risk assessment project for a K-12 school. The project can be used to accomplish tasks in the analysis and design phases in supporting the school's instructional technology. The issues related to security evaluation are also discussed by Dark (2003) who gives an overview of security evaluation, evaluation design, evaluation implementation, and evaluation management. Riola's (2006) paper discusses the issues of implementing wireless network security in the classroom. It examines security flaws in wireless networks and how wireless security is addressed in the curriculum.

This chapter will specifically focus on the security issues related to online computer labs which have to allow students to be system and network administrators.

Security Vulnerability

As mentioned before, network security failure can force an entire computer lab to shutdown. To protect the online computer lab, the first thing to do is find out the computer lab's vulnerability. The weaknesses in security may be caused by the following factors.

- Security bugs in the software.
- Insufficient security measures in preventing virus infection.
- Loose control of network access policies.
- Use of unsecured network protocols and communication ports.
- Publicly shared files and folders.
- Improperly configured application software.
- Remote access services.
- Unsecured wireless access points.
- Installation of downloaded software.

To further understand the potential vulnerability of online computer labs, let us learn more about each of the factors.

Bugs in Software

Software bugs are errors due to mistakes in the design and source code of the software. Software bugs can generate incorrect computation results or cause security flaws. The study done by the National Institute of Standards and Technology found that software errors cost the U.S. economy about $59.5 billion each year (Newman, 2002). These software bugs may be caused by:

- **Software complexity:** Newman (2002) points out that there could be millions of lines of code in a large software package. Although software developing companies spend 80% of their effort identifying and correcting defects, their software still has errors (Newman, 2002). As the size of software keeps growing, it seems impossible to eliminate all the bugs in software.
- **Insufficient testing:** Due to their market strategies, some of the software companies push their products without sufficient testing. Most of the large software companies provide the beta version of their software to let the public test the software and report the bugs. Although this cannot eliminate all the

bugs, it can significantly reduce the number of bugs. However, most of the small software companies do not provide the beta version of a product for testing. These products often cause some serious problems.

- **Lack of training and quality control:** Due to the cost and time limit, some software companies do not provide necessary training on quality controls. Employees in these companies do not have skills and motivation to prevent defects from happening in the code. Also, there may be hundreds of programmers working on a large project. It is a challenging task to detect defects caused by a large number of programmers working on the same project.

There are several things that we can do to reduce damage caused by software bugs. When selecting software for the online computer lab, choose the well-tested software if possible. Also, update the patches and service packs as soon as they are available. As for our computer lab, the software bugs will have less effect on the daily operations if the network administrator can keep updating the patches and service packs. The more serious threat is the virus infection.

Computer and Network Viruses

There are various computer viruses that can damage a computer in different ways. The following are the types of viruses and the damages that can be done by them.

- **Boot virus:** During a boot process, this type of virus overwrites or replaces the boot program stored on a floppy disk or hard disk, and then loads itself in the memory so that the computer cannot perform the normal booting. The commonly known boot viruses are Disk Killer and Stone.
- **Program virus:** When running an executable file, such as a file with the extension .BIN, .COM, .EXE, .OVL, .DRV, .DLL, or .SYS, this type of virus is loaded into memory, too. Once loaded into the memory, the virus will infect executable files that are started after the infected file. The well-known program viruses are Sunday and Cascade.
- **Trojan horse virus:** Closely related to program viruses, a Trojan horse virus is an executable file that claims it can do some helpful things such as eliminating viruses on your computer. When executed, the virus can wipe out the files on your hard disk. Some well-known Trojan horse viruses are Beast and Bifrose.
- **Stealth virus:** This type of virus infects a computer in a stealthy way. It is difficult to detect. The type of virus can add itself to a regular file and modify the file size so that the size is the same as that of the original file. In such a

way, it is hard to notice that the virus has been added to the regular file. The well-known stealth viruses are Joshi and Whale.

- **Polymorphic virus:** This type of virus can change its appearance in each different infection to make people think there may be two different types of viruses. It is also difficult to detect this type of virus. Some of the known viruses in this type are Cascade and Virus 101.

- **Macro virus:** When opening an infected word processing or spreadsheet file in which macros are created, this type of virus will be activated. If the infected file is opened on another computer, the virus will infect that computer, too. Some of the known Macro viruses are WM/Helper.C;D;E and W97M/Ekiam.

- **Java or ActiveX virus:** This type of virus is built into Web browsers. If it is enabled, it allows the currently opened Web page to take control of the computer. This kind of virus can extract any information from the computer or insert any instruction by the hacker. Some of the known viruses of this type are JS.exception and Bubbleboy.

- **E-mail virus:** This type of virus is carried by e-mail messages. It can replicate itself and send a copy to other e-mail addresses in the victim's e-mail address book. Some of the known e-mail viruses are W32/Mytob.gen@MM and Mi-mail.

- **Worm:** A worm scans a network and replicates itself in the network once it finds a security hole in the network. Unlike a computer virus which harms computer systems, a worm harms a network by consuming the bandwidth of the network. Another difference from computer virus is that a worm can propagate itself without being a part of another program like a computer virus does. Some known worms are W32/Bolgimo.worm and Supernova Worm.

Any of these viruses can do significant damage to an online computer lab. Adequate security measures must be enforced to prevent them from infecting the computers and networks.

Network Access Policies

Each device that links to a network has the potential to cause a security problem or to be infected by viruses. Network administrators have to clearly define network access policies. The devices on the network and the users who use the devices must comply with the network access policies. When setting up the online computer lab for a class, the network administrator must decide who can access what devices and use which applications, and make sure that only authorized users can access the dedicated devices. Once a user successfully logs on to a device, further access policies will be enforced to specify which servers the user can communicate with, which protocols are allowed in a communication, which network can be accessed, and which network services can be used by the user. A wrongly configured network can cause security holes. Later, we will take a closer look on network access policies.

Protocols and Ports

Many of the network protocols, such as the commonly used tabular data stream (TDS), Internet inter-ORB protocol (IIOP), and hypertext transfer protocol (HTTP), transmit information in clear text without encryption. It is easy to capture the data while the packets are transmitted over a network. A hacker can install a free network monitoring program such as Ethereal to capture the network traffic. The clear text does not stop anyone who captures data to read the transmitted information including user names and passwords. Even worse, a hacker may capture more sensitive information such as credit card numbers and bank account information from the clear text. The following are descriptions of these unsecured protocols.

- **Tabular data stream** is a protocol that is used for transmitting data between two computers. TDS was designed and developed by the Database Management System (DBMS) company Sybase and was also used by the early version of Microsoft SQL Server.
- **Hypertext transfer protocol** is a protocol used to transfer information on the Internet. HTTP was designed to publish and receive hypertext markup language (HTML) pages.
- **Internet inter-ORB protocol** is a transport protocol used for the communication between CORBA object request brokers (ORBs). Developed by the Object Management Group, IIOP is more powerful than HTTP. In addition to transmitting text over the Internet, it also enables browsers and servers to exchange integers, arrays, and more complex objects.

When used in data transmission over the Internet, these protocols do not use secure sockets layer (SSL) which uses a two-key cryptographic system. Therefore, these protocols are not secure.

There are other types of unsecured network protocols that provide services such as file transfer and network management. When these network protocols were developed, there was little or no security concern in mind. Therefore, network services supported by these protocols have no security protection. Examples of such types of protocols are file transfer protocol (FTP), Telnet, and simple network management protocol (SNMP). The following are descriptions of these protocols.

- **File transfer protocol:** FTP is a protocol used to transfer entire files between two computers over a network. An anonymous FTP does not require a user name and password for login. Therefore, it is not secure. Even if it is not anonymous, FTP is still not secure. For example, the FTP service permits an unlimited number of attempts to enter the FTP service. This allows a hacker to use a brute force "password guessing" scheme to enter the FTP service.

- **Telnet:** Telnet is an Internet protocol used to connect to a remote computer. It actually allows a user to remotely log on to a server computer. Among many security problems, the most vulnerable one is that Telnet transmits information in clear text. Through Telnet, a user who has an account on the host computer can gain more power than what has been assigned to him/her. The Telnet service can be crashed by a denial-of-service (DoS) attack which consumes the bandwidth of the victim network or overloads the computational resources to cause the loss of the network service.

- **Simple network management protocol:** SNMP is a network management protocol used to monitor and control network devices and to manage configurations, statistic collection, and performance. SNMP can query and manage configuration settings on network devices such as computers and routers. However, SNMP does not provide a facility for log on authorization. This can open the door for hackers to take control of network devices.

In fact, most of the network protocols, such as TCP/IP, were not designed with security in mind. These protocols can all be considered as unsecured protocols. To be a secure protocol, the data transmitted must be encrypted such as HTTPS which is a combination of a normal HTTP interaction over an encrypted secure socket layer. We will learn more about HTTPS later.

Ports that are used by unsecured protocols are unsecured ports. For example, the port 80 used by HTTP is an unsecured port. Even though some of these ports are potentially unsafe, we may have to keep them open during lab operations, such as the port 80 used for the Internet connection, and find other ways to protect the computer lab.

Shared Files and Folders

File sharing services can expose files on a computer to other users. Shared files and folders can potentially make a computer system unstable or insecure. When a computer hosts a shared file, a shared folder, or a shared drive, it will allow, by default, a user with a guest account to access the computer. By taking advantage of this fact, certain worms can propagate by using the shared hard disks. Sometimes, the files created on a computer may contain the information about the computer and the creator of the files. Through the shared files, this type of information may leak out.

Application Software

When we install application software such as a Web browser, e-mail, or database management system and make it available on the Internet, we are running the risk of exposing the host computer and the network to hackers. The following are some of the application software that is commonly attacked.

- A Web browser is probably the most vulnerable software to be attacked by various viruses. Interactive services such as ActiveX controls and Java applets can potentially allow viruses or other malicious software to enter the computer system. Through ActiveX controls and Java applets, the malicious software can even bypass a firewall to get into the private network. Some of the Web browsers can automatically open a Word file stored on a Web server. If the Word file contains malicious code, the user's computer system will be infected.

- Web servers allow scripts to be executed on the server side. A script installed on a Web server may contain bugs; this is a potential security hole. A Web server can potentially expose the private network to the entire Internet. Through a Web server, a hacker can cause damage ranging from altering the content on the Web site to taking away the entire database of sensitive information.

- A lot of application software programs such as the products included in Microsoft Office contain macro functionalities. Many macro viruses are designed to attack this type of software. Some Microsoft Office products may suffer from the remote code execution attack which allows remote attackers to execute code through the Windows shell.

- E-mail is another application that is vulnerable to attackers. In several ways, an attacker can damage an e-mail system. Attackers may set up a news server. When accessed, the news server will send malicious responses to overflow e-mail clients. Attackers may also send some malicious document in an e-mail message. When opening it, the user's computer will be infected. E-mail spam is another security threat. A spam sends replicated messages to thousands of recipients.

- Database is also application software that is often attacked by hackers. If database servers allow users to bypass the operating system to directly access its database, this can open the door for attackers. A database may suffer from attacks such as attacks of all its TCP/IP communication ports so that the buffer listening to these ports overflow with malicious data. This can lead to overwriting adjacent memory locations and causing the host computer to crash. The application tools of a database may also be convinced to run malicious SQL code which will cause some significant damage. Some of the DBMSs have well-known default administrator accounts with a standard or blank user name and password. If not reconfigured after installation, it will leave a big security hole to allow intruders to take full control over the devices or programs.

- The widely installed media player is another weak spot in computer security. Through a media player, an attacker can damage a computer without requiring much user interaction. When a user surfs a malicious Web site, malicious software such as spyware and Trojan horses can also be introduced to the user's computer system through the media player.

- Instant messaging is a communication tool used for collaboration. It allows live communication through the Internet. Participants know whether a peer is available through an online status indicator. This type of application communicates through Internet relay chat (IRC) which is a form of instant communication channel. IRC is presenting an increasing security threat. Instant messaging can cause buffer overflow that may crash a user's computer system. Also, instant messaging allows file transfer over a network. During file transmission, worms can sneak in and replicate themselves on the network through an IRC channel. IRC channels can also be used for launching a denial-of-service (DoS) attack.

As you can see, almost all of the security problems are caused by Internet connections and by opening malicious files. Later, we will discuss how to deal with these problems.

Remote Access

Remote access is crucial for an online computer lab. However, remote access can also generate some security concerns. Many of the server-side services such as Terminal Services, virtual private network (VPN), and virtual machines allow students to remotely access the host server. Sometimes, these services require a local account for the user to log on. Also, some application software such as database servers can also allow remote access. The following are some of the vulnerabilities of remote access.

- Weak authentication is one of the major security problems. Remote access servers often have default log on configurations that may use well-known user names and passwords, or simply have a blank password. Hackers can use these standard user accounts to create administrator accounts for themselves before these standard accounts are changed.

- The remote access service may also be subject to attacks such as remote code execution on the host server or a denial-of-service (DoS) attack.

- virtual network computing (VNC) can be used to remotely control another computer. Over the network, VNC can transmit keyboard presses and mouse clicks from one computer to another. The data stream transferred by VNC is not encrypted, which may be a potential security problem.

Wireless Network

Computers in a wireless network are linked by radio waves. This method of communication allows anyone within the radio range to be able to use the network for data transmission. This means that a hacker can sniff the network without being physically wired to the network. Some of the commonly seen wireless network security problems are discussed below.

- The security problem of a wireless network starts from the installation of wireless network equipment. In many wireless network products, the security measures are turned off by default for quick installation. If not properly reconfigured, a wireless network has no authentication protection. Anyone within the radio range can use the network. An unauthorized wireless network access point can be misused by some users. For example, a user may broadcast a large amount of data or even spread viruses through the network.

- Another security problem is that, when a computer is booted up, it will send a broadcasting message to find an access point nearby. If an attacker in the radio range captures the broadcasting message, the attacker can use the same information to create a network connection. Once the attacker is linked to the victim computer, it is possible for the attacker to access the files on the computer's hard disk.

- An attacker can set a rogue access point. When a computer is turned on, the rogue wireless access point will offer to connect to the victim computer. Once connected, all the traffic that the user enters, including the user's user name, password, and credit card information will go through the attacker's rogue access point and will be sorted out with packet sniffing software.

- Many publicly available wireless access points are deployed without having the encryption ability. It is very dangerous to transmit unencrypted data over a publicly available network. Even with the encryption turned on, some of the encryptions are not secure enough for a wireless network. The widely used 64-bit encryption wired equivalent privacy (WEP) used by the 802.11b wireless standard is a weak encryption. The encryption can be easily cracked with free-download software.

When students use a wireless network at home, it is possible that they can expose their log on information to the public. If not properly configured, a wireless computer lab may become available to the public. Even if the lab is properly configured, students may reconfigure the settings later since they practice being the administrator. Later, we will discuss the protection of wireless computer networks.

Downloaded Software

Installing downloaded files can cause security problems. Viruses such as Trojan horses can hide in an executable file pretending to help you remove viruses from your computer. Once the file is downloaded and installed, the computer will be infected. Spyware is another security problem that can be introduced by downloading software. Spyware is a kind of script running on the victim computer to secretly collect information about the user and the computer. Then, the collected information is secretly transmitted to a hacker's Web site or e-mail address. Spyware can spread itself by the following means.

- Get Internet surfers to click a malicious link that will trigger the installation of spyware.
- Get users to interact with a pop-up ad. A user is often prompted to accept a certain condition. Once the user accepts it, his/her computer will be infected.
- When a user installs downloaded free software.
- Use a drive-by download which automatically downloads spyware to a computer without the user's knowledge.
- Spyware can also be distributed through e-mail.

During hands-on practice, students need to unblock some interactive programs such as ActiveX, VBS script, JavaScript, and Java applets for remote access. They also need to download some files from other Web sites. These will make the online computer lab and students' home computers very vulnerable to spyware and viruses.

Now, we have discussed various kinds of security problems. The key question is how we can prevent these bad things from happening. The first step is to assess the security vulnerability in a network and the computer systems connected to the network. Then, we will be able to find a way to deal with it. Let us start with security assessment.

Security Assessment

With such a variety of vulnerabilities, there is an urgent need to identify the security holes in an online computer lab and students' computers. To accomplish this task, we need to perform the following tasks.

- Create a checklist of all the items that need to be assessed by the security assessment process.
- Select security assessment tools that can get the job done.
- Classify and document the security vulnerabilities based on the threat level.

The findings through these steps can be used as a guideline for making security policies and for enforcing security measures. In the next a few paragraphs, we will discuss the issues related to each of above tasks in detail.

Security Assessment Checklist

The first thing to do in a security assessment process is to identify a list of things that need to be checked. We need to find out what to evaluate. We may need to evaluate vulnerabilities from the areas: Computer systems, network systems, remote access, and application software. The checklist may be different for different computing environments. The following are some sample checklists of vulnerabilities in different areas.

Checklist for Computer Systems

For a computer system, a security assessment process should check the following items that may result in potential vulnerability.

- If an administrator account has a strong password.
- If there are unnecessary services turned on.
- If there are unnecessary accounts.
- If there are shared files, directories, or disks.
- If there are guest accounts.
- If anonymous access is allowed.
- If a user is granted more access power than he/she needs.

- If there is an account lockout policy.
- If security event auditing is enabled.
- If antivirus software is installed and updated.
- If service packs and critical patches are installed and updated.

Checklist for Network Systems

For a network system, the following items should be checked during a security assessment process.

- If routers and switches are updated with the latest patches.
- If routers are required to have a strong password, enabled ingress filters, egress filters, security notification services, and auditing services.
- If there are vulnerable ports.
- If routers and switches disable unused services and Web-facing administration.
- If routers' interfaces are secured and network traffic is encrypted.
- If firewall filters are in place, unused ports and protocols are blocked by default, and intrusion detection is enabled.
- If the firewall enables IPsec and is configured for securing packet flows and key exchange.

Checklist for Web Servers and Remote Access

The following is a sample checklist for Web servers and remote access.

- If the Web server is updated with the latest service packs and patches.
- If the services such as ASP.NET state service, Telnet, FTP, SMTP, and NNTP are disabled by default.
- If the protocol WebDAV is disabled when not used, TCP/IP stack is hardened, and NetBIOS and SMB are disabled.
- If unused accounts, anonymous accounts, and guest accounts are removed.
- If the default accounts have been reconfigured and the Everyone group is restricted.
- If there is a strong password policy.

- If Web site content files and log files are placed on different non-system volumes.
- If there are shared files, directories, or disks.
- If there are unsecured ports, log files, paths, Web sites, scripts, and virtual directories.
- If the remote registry access is restricted.
- If there is log-on auditing services.
- If the certificate and its public key are valid.
- If permissions for code access are restricted.
- If security tools are installed.

Checklist for Databases

Application software such as a database management system (DBMS) allows remote access through the Internet, which may cause some security vulnerability, too. The following is a sample checklist for a DBMS.

- If the DBMS is upgraded with the latest security packs, updates, and patches.
- If unused and unnecessary services are disabled.
- If the database is installed on a different computer from domain servers.
- If least-privilege accounts are assigned to users for services that have to be on all the time.
- If the DBMS enables the remote log on that bypasses the operating system's authentication.
- If there is a strong password policy.
- If there are unnecessary protocols and ports that need to be turned off.
- If there are unused user accounts and guest accounts that need be removed.
- If the default accounts are reconfigured.
- If the data files and log files are secured.
- If sensitive data are encrypted.
- If there are data stored in shared files, folder, and disks.
- If failed activities are logged and the log files are secured.
- If the database auditing service is enabled.
- If permissions for accessing database objects are properly configured.
- If the access to sample databases and procedures are restricted.

To keep a computing environment secure, security assessments should be regularly performed. Like the above sample checklists, you can create your own checklists. You can manually check each of the items on a list, or you can use a security assessment tool to help you get some of checking done.

Security Assessment Tools

There are many security assessment tools available. Most of these tools are freeware and very simple to configure. Some of these tools are platform independent online-based tools. To help you choose the right tools for a specific security assessment job, the descriptions of some commonly used tools are given in the following.

- **Basic Port Scanner**. This type of tool can be used to search open ports and collect information on currently active services on a network. Some of these tools are designed specifically for Windows operating systems and some of them are designed for Linux operating systems. You may find many platform-free online-based port scanners that allow the user to operate a port scanning process online.

- **Web scanner**. This type of scanner tests the configuration of the Web server and searches for scripts with known vulnerabilities. A Web scanner can also be used to test the vulnerability in remote log in by using brute force account testing to see if the password is too easy to guess or if the password is a default password.

- **Virus scanner**. This type of scanner can be used to scan files and warn the user if a file is likely to contain a virus. This kind of tool is capable of detecting malicious software in a computer's random access memory (RAM) or hard drive. There are a broad range of such tools available for free download. These tools are designed to detect and remove viruses such as dialers, Trojans, and many other kinds of malware.

- **Firewall scanner**. This type of scanner provides comprehensive security auditing. It tests the configuration of a firewall and reports to the service enabled through the firewall. It can test for both proxy-based and filter-based firewalls. It can identify the level of vulnerability and find balance between usability and security.

There are also some comprehensive security assessment tools which can find the security vulnerability in multiple areas. These tools can be used to analyze the entire security architecture for an enterprise by comparing the best practice commonly used in an industry. Typically, these assessment tools will scan a computer or a network system. The result of the scanning will be used to compare with the known vulnerability stored in a database. If there is a problem, the security assessment tools will produce a report. Based on the design specifications, we may have the following types of comprehensive security assessment tools.

- **Assessment tools for large enterprise:** This type of security assessment tool can be used to scan a large distributed network system, and it allows the user to run a checklist to find out which components in the network are at risk. It provides the latest vulnerability information and threat correlation. It also allows the user to determine the severity of a threat so that critical vulnerabilities can be fixed first.

- **Assessment tools for small- or medium-sized business:** These security assessment tools are designed to assess security problems in a small or medium-sized network. They can be used to detect security misconfigurations and missing security updates on a computer system. To help users make decisions on security measures, some of the tools can analyze the security status and give remediation guidance.

- **Analytical assessment tools:** This type of security assessment tool is a more sophisticated security assessment tool that is designed to assess weaknesses for a medium-sized enterprise. It will first ask the user to answer a series of questions on a checklist related to various topics. Once the tool gets all the answers, based on the information collected, it conducts assessments. Based on the results of the assessments, it will provide a list of recommendations for solving potential security problems.

- **Demand subscription assessment tools:** This type of assessment tool provides on-demand subscription services for vulnerability management and policy compliance. It provides a hardware appliance installed inside the firewall to scan any device with an IP address and analyze data collected from the scanning. There is no software to install. With analytical results, it can efficiently strengthen an enterprise's security by complying with the internal policies and external regulations.

- **Assessment tools for detecting software bugs:** These security assessment tools are used to detect bugs in software. They provide Collaborative Development Lifecycle Management solutions for software development teams. They track defects in a multilevel role-based process. Some of the tools are designed for high-level project management. Some of them provide the online services for tracking defects for software development. It provides a function-rich defect-tracking system, which is a great benefit for small companies. Many open-source defect-tracking tools are also available such as those J2EE-based defect-tracking and project management software designed to make team work easier.

These security assessment tools provide great help in identifying security problems. Especially, for a medium or large enterprise, it is impossible to accomplish all these security assessment tasks by hand. To efficiently deal with security problems, we need to classify the scanning results, discussed in the next paragraph.

Classifying Security Vulnerabilities

A typical security vulnerability scan may return thousands of findings. To effectively analyze these findings, it is important to classify the assessment results. It is necessary to understand that the damages may be caused by security problems and to identify the problems that matter the most. Based on the chance of a computer lab being attacked and the potential impact of an attack, the potential security vulnerabilities can be classified into different levels such as critical, serious, moderate, and minimal. In the following, let us see how each of the levels is defined.

- **Critical level:** If the security vulnerability allows an attacker to run malicious code on the victim computer system without requiring a user's action to trigger the execution, it can easily crash the system or even wipe out the entire hard disk(s). This type of security vulnerability is at the critical level. Software bugs and misconfigured operating systems and special purpose servers can also cause this type of problem. Once this type of security vulnerability is identified, immediate action is required to fix the problem. A warning about the problem should be sent out immediately.

- **Serious level:** If the security vulnerability allows unauthorized users to get higher privileges to view the information that these users are not supposed to read or the unauthorized users gain higher privileges to stop certain services, the system's confidentiality, integrity, or availability of resources will be hurt. This type of security vulnerability is at the serious level. For example, by clicking an e-mail attachment, a worm may be introduced to the system. The worm will consume the bandwidth of a network and eventually stop some of the network services. Again, a serious level security problem should be fixed right way, and a warning should be sent out immediately.

- **Moderate level:** If the security vulnerability allows spyware to collect information about the Web sites you have been visiting, or a new virus has been reported somewhere else and has the potential to spread to your network system, this type of security vulnerability has the potential to harm your system. Even though the damage may not happen now, it can happen at anytime if the right circumstances are formed. The user is still required to take quick actions to prevent any future damage. At the same time a warning message should be passed to all the users immediately.

- **Minimal level:** For this level, no unusual activity is carried out beyond the routine check for security vulnerability. This level causes minimal consequences in damage. You only need to make the general security education available to the users. No immediate action is needed and no warning needs to be sent out.

Based on the levels of severity and the recommended activities, system managers and network managers can decide how to react to a security problem. The decision on how to enforce security measures to solve a problem identified by a security assessment process will depend on the policies set by the joint effort of administrators, technicians, and users. Next, we will discuss some issues related to security policies.

Security Policies

As we all know that there is no absolute secured system. To reduce security problems, the fundamental step is to decide what to protect. That is, we need security policies. Creating security polices should be done by a team including the network manager, the institution's leaders, and the developers so that the policies can reflect suggestions from the people who are involved in the computer lab construction. The following are factors that need to be considered when drafting security policies:

- The university's regulations and procedures.
- The security practice by the computer service department.
- The cost factor, the value of the information being protected, and the level of technical difficulty.
- The impact on the computer lab's usability, scalability, performance, and maintenance.

The security assessment process will provide a list of vulnerabilities. The security policies will prioritize the vulnerabilities and provide a guideline for dealing with them. The team should come up with an agreement on what to protect and how to handle the vulnerabilities. The security policies should regard the following three areas:

- **Confidentiality:** Prevent the information from being viewed by unauthorized individuals.
- **Integrity:** Prevent the information from being intentionally or unintentionally altered.

- **Availability:** Make sure that the information is available to the authorized users when needed.

Confidentiality

For confidentiality, a list of rules should be specified in the security policies for the following concerns:

- Who should access what information? For example, a student is allowed to access all the data stored in a database created by him/her but not other students' data.
- Who should access which computer? For example, students are allowed to access the lab VPN server.
- Who should be allowed to do what? For example, students can be the administrators of their group servers in an online computer lab.
- What data should be encrypted?
- How do we define a strong password policy?
- How do we monitor the network system? Specify the system monitoring procedures and technologies.
- What is the damage if sensitive information is viewed by an unauthorized person? For example, the damage can be serious if a student's user name and password is viewed by others.

Integrity

For integrity, rules should be specified to answer the following questions:

- Who should be allowed to alter what information? Students are allowed to modify the configuration of their own server but not that of the VPN server.
- How do we avoid protocol related security problems? For example, use HTTPS to transmit encrypted data over the Internet.
- How do we prevent viruses? For example, update antivirus software as soon as an update is available.
- How do we prevent hackers from altering or stealing sensitive information? For example, turn on encryption for wireless access points.

Availability

For availability, the policies should specify rules for the following.

- The time frame that certain information should be available to qualified users. For example, teaching materials should be available to students 24 hours a day, 7 days a week.
- The severity of damage due to unavailable service or information. For example, if the online computer lab is unavailable for a few hours, it may not cause serious damage. However, if it is not available for a few days, the problem can be serious.
- The methods of accessing network to allow qualified users to get information when needed.

The security policies should be precise and compact. They only need to indicate what to do for the questions related to confidentiality, integrity, and availability. The policies will serve as a guideline on:

- What is and is not permitted on the network.
- Which technology to use to solve which security problem.
- Who should use which device.

After the security policies are created, educating the computer lab users about the policies is a crucial step for successfully carrying out the policies. One cannot expect the students and faculty members to automatically comply with the policies. They need to understand the policies and know how to cooperate with the computer service department. To gain support from the university's administrators, it is also important to educate them.

After the security policies are created, they should be monitored, updated, and well documented. Over time, the university's regulations may change, new threats may pop up, and the computer lab may be upgraded with new technologies. The security policies should keep up with these changes. Well-documented security policies will make the task of policy modification much easier.

Security Management

After the security policies are created, the next step is to enforce the policies with various security measures.

Facing a list of security vulnerabilities, decisions need to be made on how to deal with them. The following are three different treatments.

- Perform routine security management as usual if the threat level is low.
- Send a warning message to users about the potential problem.
- Enforce security measures immediately if the threat level is high.

In the following, we will look at some of the tasks that will be performed in each treatment.

Routine Security Management

There are two types of routine security management tasks. The first type does not require regular activities. It is only performed based on demand such as creating user accounts and installing new devices. The other type of security management task is performed regularly in a certain time period such as a routine security assessment.

The security management based on demand may perform tasks in the areas of computer system, network system, remote access, and other applications. The following are some sample security measures performed in each area.

Computer System Security Management

Many of the critical security problems are caused by computer systems. The weaker security measures on a computer system can invite attackers to run malicious code on it. Very carefully follow the security policies and make sure that a new computer system is secure before you link it to a network. The security management tasks to be performed on a computer system may include the following:

- Install an operating system by using a secured file system.
- Create user accounts and assign privileges to the users based on the security policies.
- Use strong passwords for the user accounts.

- Remove the unused and unnecessary user accounts including the unnecessary default accounts and guest accounts.
- Turn off the unused and unnecessary services such as Telnet and anonymous access.
- Turn on the automatic security service.
- Restrict the access to the shared files, folders, and disks.
- Set the account lockout policy.
- Install the latest security packs, patches, and antivirus software.
- Distribute the latest security packs, patches, and antivirus software to students and faculty members.
- Enable the security event auditing service and the log warning service.
- After the installation of a software package or a new hardware device, run security assessment software to see if there is security vulnerability.

Not only do network administrators need to carefully enforce these security measures, users including students and faculty members also need to do the same. A list of security measure enforcement tasks should be available to everyone who is involved in an online computer lab project. Fix all the problems before you link computers to a network.

Network System Security Management

A network is another most vulnerable place that can cause disastrous damage. Once the network is infected, all the computers and network devices connected to the network will be affected. Extra caution needs to be taken before the network is deployed and used by the public.

- Install the latest patches on routers and switches.
- Enable the security filters, monitoring service, notification service, and auditing service.
- Disable the unused services.
- Use strong passwords.
- Block all the vulnerable ports.
- Secure routers and switches.
- Encrypt the network traffic.
- Block the unused protocols by default.

- Configure the network not to receive or forward broadcast messages.
- Enable the firewall and configure it according to the security policies.
- Update the firewall with the latest patches.
- Enable IPsec on the firewall and configure it for encrypted communication.
- After the network equipment is installed, run security assessment software to see if there is security vulnerability.

The network security measures should be fully enforced while creating the network. Most of these tasks should be done by network administrators who have the security management experience and the knowledge.

Web Servers and Remote Access

A Web server and remote access service are the first defense line to protect your network system. If the Web server or remote access service is not properly configured, it is likely to open the door for attackers to come to your network. In fact, most of the security problems are caused by Web servers. A Web server and remote access service must be properly configured before they can be used on the Internet. The following are some security measures for Web servers:

- Install a Web server software package on a dedicated computer.
- Web sites should be created on a secured file system.
- Only allow the administrator of the host server to log on to the server locally, and limit the number of administrator accounts. One or two such accounts should be enough.
- Update the Web server with the latest service packs and patches.
- Disable the unsecured services such as ASP.NET state service, Telnet, FTP, SMTP, NNTP, NetBIOS, WebDAV, and SMB by default.
- Harden the in-use protocols such as TCP/IP.
- Remove the unused accounts, anonymous accounts, and guest accounts from the Web server.
- Reconfigure the default accounts and use strong passwords.
- Restrict the Everyone group by preventing some members of the group from accessing certain services.
- Place the Web site content files, log files, and system files on different volumes.
- Restrict access to the shared files, directories, or disks.

- Block access to unsecured ports.
- Restrict access to log files, scripts, paths, Web sites, and virtual directories.
- Restrict the remote log-on and registry remote access.
- Enable site monitoring, system file access auditing, and log-on auditing.
- Make sure that the certificate and its public key are valid.
- Remove the Web server's sample applications.
- After the Web server and remote access service are installed, run security assessment software to see if there are security vulnerabilities.

Once the Web server is deployed, the internal network is connected to the external Internet and the risk to the computer lab is increased dramatically. Make sure that all these security measures comply before the Web server is deployed for public access.

Special Purpose Servers and Remote Access

The application software is also a path from the external network to the internal network. The security risk for Internet accessible application software is also very high. Like Web servers, if special purpose servers are not properly configured, they can open the window for hackers to peek through. A database management system is a typical type of such application software. Both Microsoft SQL Server and Oracle DBMS allow remote access through the Internet. Make these applications fully comply with the security policies before they can be used on the Internet.

- Install the database server on a dedicated computer.
- Data files, log files, and system files should be secured and stored on different volumes.
- If it is possible, use a more secure operating system account to log on to the database server.
- The network services such as TCP/IP need to be hardened.
- Make sure that the default accounts are reconfigured and strong passwords are used.
- Update the DBMS with the latest security packs, updates, and patches.
- Disable the unused and unnecessary services.
- For the service that needs to be on all the time, assign the least privilege to the user.
- Turn off the unnecessary protocols and ports.

- Remove the unused user accounts and guest accounts.
- If necessary, encrypt the sensitive data.
- Restrict access to the shared files, folders, and disks.
- Restrict the remote log-on to the database server.
- Enable the monitoring, auditing, and logging services.
- Configure the permissions for accessing database objects according to the security policies. No permission is given to the public.
- Restrict the access to sample databases and procedures.
- After the database management system is installed, run security assessment software to see if there are security vulnerabilities.

Normally, the security measures on a special purpose server are weaker than those for an operating system. Therefore, special purpose servers are often the target of attackers. We must pay attention to the security vulnerabilities of a special purpose server and fix the problems as soon as they are identified. For first-time installation, make sure that all the security measures are enforced before the special purpose server is deployed for public access.

Also, it is important to have a maintenance plan that includes the schedule and tasks to be performed. The following are some tasks commonly carried out in regular maintenance.

- Regularly perform security assessments and find answers to the questions in the security assessment process.
- Update the latest service packs, patches, and antivirus software as soon as they are available.
- Inform the related people about possible threats and security vulnerabilities.
- Review and analyze monitoring reports, log files, and auditing results.
- Back up database files regularly.
- If security vulnerabilities are found, fix the problems.

The installation of antivirus software can dramatically reduce the security risk of an online computer lab. Usually, a university has a package deal with a software vendor. However, some students may not know the availability of such software. When students use their home computers as client computers, make sure that every client computer has antivirus software installed. The commonly used antivirus software packages have the following features:

- They include antivirus, antispyware, antiworms, antispam, and firewall tools.
- They have scan tools that can automatically search for viruses.
- Some of the antivirus software packages can be subscribed by users. They allow users to upgrade their antivirus software through the life of their subscription.

It is better to perform regular maintenance before a semester start. At this time, many new devices are installed and software is upgraded. The lab manager needs to make sure that these hardware devices and software upgrades are properly secured. If anything needs to be fixed, it will have a minimal impact on the classes. Another good time to perform security maintenance is before final exams. This will make the computers and network systems function properly so that students can get their assignments done before the final exams.

Distributing Warning Messages

Based on the nature of a warning message, the message can be delivered in several ways. It can be delivered to an individual, to a group of users, or to a whole campus.

- If a security assessment process has found a security vulnerability on a single computer system, the security warning message can be sent to the owner of that computer and ask the owner to fix the problem immediately. The message can be delivered through e-mail or a telephone call.
- If a security assessment process has found security vulnerabilities in a local area network or a computer lab, the warning message should be sent to the users of the network or the computer lab. E-mail is a more efficient way to deliver the warning message.
- If a security vulnerability impacts an entire campus, or a security warning is issued by the computer service department, or warnings come from hardware and software vendors and they have a wide impact, a warning message should be sent to the whole campus. Even if students may use their home computers as client-side computers, the warning message should still be sent to the students through e-mail.

Warning messages should be sent as soon as they are available. It is even more efficient if the warning message is attached with information on how to fix the security problem and how to prevent the same problem from happening again.

Emergency Response to Security Vulnerabilities

If the security policies are not well developed and the security measures are not rigorously enforced, damage to the computer lab will happen eventually. As described earlier in this chapter, the damage of software bugs, incorrect configuration, insufficient security measures, and mal-operations may cause the following consequences:

- Corrupt files.
- Wipe out files on disks.
- Steal sensitive information
- Make services unavailable.
- Block network traffic.
- Slow down system and application performance.
- Shut down computer systems.

Once these problems happen, the situation can cause panic among the users. To better prepare for actual response and recovery of a computer lab, it is wise for a computer service department to develop an emergency response plan. In the plan, the following issues should be addressed:

- Identify the individuals who are responsible of specific tasks.
- Identify the resources that might be needed in a crisis.
- Train these individuals with necessary skills for disaster recovery.

During a security crisis, the emergency response team will perform the following activities.

- Identify the cause of the problem.
- If necessary, disconnect the computer lab from the Internet.
- Clean up the virus infection or reconfigure the computers or network to remove the security problem.
- Revise the current security policies to prevent future security problems.
- If necessary, reinstall all the software, and reconfigure all the services according to the new security policies.
- Run security assessment and fix all the problems found by the assessment.

- Educate the users as to what they should do and not do.
- Reconnect the computer lab to the network.

Every effort made in an emergency response plan will be paid off during a security crisis. With training, individuals will learn how to recover the computer lab quickly under stress. Of course, the best way to handle a security crisis is that it will never happen. Therefore, well developed security policies, rigorously enforced security measures, and carefully conducted security assessments are the best security practice for online computer labs.

Conclusion

This chapter has covered the security related issues. It has discussed topics such as security vulnerabilities, security assessments, security policies, and security management for an online computer lab. To understand the sources and the damage of security problems, this chapter first described various security vulnerabilities which may be caused by software bugs, weak access controls, wrong configurations, virus infections, and so on. This chapter has discussed security assessments in the areas of computer systems, network systems, Web servers, remote access, and special purpose servers such as database management systems. It has also introduced several commonly used security assessment tools. Then, this chapter discussed the issues related to security policies. It looked into the factors that may influence security policies. It then listed the issues that needed to be taken care of in the areas of confidentiality, integrity, and availability.

About the topic of security management, the chapter has discussed three aspects, security management, distribution of warning messages, and emergency response to a security problem. For security management, this chapter has provided a sample list of security measures to be enforced in the areas of computer systems, network systems, Web servers, remote access, and database management systems. It then discussed the issues related to sending warning messages. Regarding to the aspect of emergency response, the chapter has discussed how to prepare for a security crisis and has given a list of things to do when a disaster happens.

Security vulnerabilities can cause serious damage ranging from reducing a network's performance to shutting down the entire Web-based teaching system. To successfully implement online computer labs, the security issues must be addressed and properly resolved. After the security issues are taken care of, our next task is to develop lab-based teaching materials. The development of teaching materials will be covered in the next chapter.

References

Barman, S. (2001). *Writing information security policies.* Indianapolis, IN: Sams.

Clercq, Jan De. (2004). *Windows Server 2003 security infrastructures: Core security features.* Oxford: Digital Press.

Dark, M. J. (2003). Evaluation theory and practice as applied to security education. In C. Ervin & H. Armstrong (Eds.), *Security education and critical infrastructures* (pp. 197-214). Norwell, MA: Kluwer Academic Publishers.

El-Khatib, K., Korba, L., Xu, Y., & Yee, G. (2003). Privacy and security in e-learning. *International Journal of Distance Education, 1*(4), 1-16. Retrieved April 7, 2007, from http://iit-iti.nrc-cnrc.gc.ca/publications/nrc-45786_e.html

Fung, K. T. (2004). *Network security technologies* (2nd ed.). New York: Auerbach.

Kaeo, M. (2003). *Designing network security* (2nd ed.). Indianapolis, IN: Cisco Press.

Kaufmann, R., & English, B. (2004). *MCSA/MCSE: Windows Server 2003 Network Security Administration study guide (70-299).* Alameda, CA: Sybex.

McNab, C. (2004). *Network security assessment: Know your network.* Sebastopol, CA: O'Reilly Media, Inc.

Newman, M. (2002). Software errors cost U.S. economy $59.5 billion annually. NIST News Release. Retrieved April 7, 2007, from http://www.nist.gov/public_affairs/releases/n02-10.htm

Ortmeier, P. J. (2004). *Security management: An introduction* (2nd ed.). Upper Saddle River, NJ: Prentice Hall.

Peltier, T. R. (2004). *Information security policies and procedures: A practitioner's reference* (2nd ed.). New York: Auerbach.

Riola, P. (2006). Wireless security in the classroom. In C. Crawford et al. (Eds.), *Proceedings of Society for Information Technology and Teacher Education International Conference 2006* (pp. 1027-1032). Chesapeake, VA: AACE.

Smith, P. G. (2005). *Linux network security* (Administrator's Advantage Series). Hingham, MA: Charles River Media.

Turnbull, J. (2005). *Hardening Linux.* Berkeley, CA: Apress.

Weippl, E. R. (2005). *Security in e-learning* (Advances in Information Security). New York: Springer.

Yang, D., & Dark, M. (2005). A service learning project of information security risk assessment for K12 school corporations. In C. Crawford et al. (Eds.), *Proceedings of Society for Information Technology and Teacher Education International Conference 2005* (pp. 1672-1677). Chesapeake, VA: AACE.

Chapter IX

Developing Online Computer Lab-Based Teaching Materials

Introduction

To get an online computer lab ready for teaching and hands-on practice, instructors should develop lab manuals and other online teaching materials. The lab teaching materials can be text, a combination of text and figures, or truly multimedia based materials including figures, sounds, animations, and video clips. By including multimedia in the instruction, the quality of the lab teaching materials can be greatly improved. The use of technology can make the tasks of developing multimedia teaching materials relatively easy. On the other hand, the use of multimedia teaching materials is limited to the type of hardware and software that are available on the server and client sides. Also, different instructions will have different requirements for multimedia materials. In this chapter, we will take a closer look at the issues related to the design of lab instructions and multimedia based teaching materials.

We will first take a look at the design of lab-based teaching materials. We will discuss how to design the teaching materials that meet the special needs of lab instructions.

Unlike other online teaching, lab-based teaching requires the illustration of hands-on activities. It also needs to give clear instruction on how to involve students in collaborative activities. Students should also be given information about security issues. The lab teaching materials should also include instructions on the use of multimedia technologies for lab activities.

After we have investigated the issues related to the design of lab-based teaching materials, we will further discuss the topics related to the development of multimedia based instructions. Demonstrations of some general approaches will be given on developing multimedia based instructions, including the software and hardware used by these approaches. We will examine how to use multimedia tools to create multimedia teaching materials. There will be some information on how to record sound and video, deploy multimedia files, present multimedia teaching materials online, develop a collaborative environment, and capture screenshots for hands-on practice.

The last part of this chapter will show a case study of developing lab-based teaching materials. To illustrate how the lab teaching materials can be used in a lab-based online teaching class, several examples will be used to demonstrate the process of developing lab-based teaching materials. Some of them have simple text-based lab manuals and some contain multimedia based demonstrations. The examples will also be used to illustrate the online teaching materials used by different courses such as network management and database system development.

Background

As e-learning is getting more prevalent in computer based training and computer based teaching, e-learning instruction has become one of the widely studied research topics. Instructors need to resolve many common problems encountered in e-learning instruction. To create better e-learning course materials, Web-based instruction must be carefully prepared. Instructors need to decide how to present course materials online, how to make the course content easier to understand, and how to properly use multimedia course content (Clark & Mayer, 2003).

To assist the design of Web-based teaching, instruction models such as the ASSIST-ME Model have been developed (Koontz, Li, & Compora, 2006). The design and implementation of Web-based teaching materials needs to use technologies. Instructors should have adequate training on using these technologies. This is particularly important for developing lab-based teaching materials of technology-based courses since the content of the teaching materials is closely related to IT products on the market. The technologies used to develop these teaching materials must be able to demonstrate the use of the IT products (Kruse & Keil, 2000).

Developing multimedia course content for an online computer lab needs various audio and video content development tools. These tools can be those sold by hardware and software vendors, or they can be those included in a learning management system (LMS). It is very convenient to use the tools provided by a LMS package if a university has already purchased the package such as WebCT (Rehberg, Ferguson, McQuillan, Eneman, & Stanton, 2004), Blackboard (Southworth, Cakici, Vovides, & Zvacek, 2006), and Moodle (Cole, 2005). The tasks, such as setting up online classes, deploying multimedia materials, collaborating with students, assigning homework, giving quizzes, organizing forums and chat rooms, creating journals and surveys, and evaluating students' performance can all be accomplished by using the tools provided by these major LMS packages.

Sometimes the tools provided by a LMS do not meet all the needs for developing multimedia based teaching materials. In such a case, you may consider using more sophisticated multimedia development tools such as Macromedia Studio (Bardzell, Bardzell & Flynn, 2005). Developing multimedia course content often involves multiple multimedia development tools such as Macromedia Breeze to build dynamic multimedia Web projects (Adobe, 2006). Instructors should have adequate training on using these tools.

Web-based interactive teaching and learning can be accomplished through the technologies used in the gaming industry (Watkins, 2005). Game-based activities can help trainers and instructors make an online class more dynamic and exciting. Interactivities can also be accomplished by using collaboration tools such as chat rooms and discussion boards. Much research has been done on implementing collaboration. Researchers and practitioners found that collaborative learning with technology can greatly improve teaching and learning (Roberts, 2004). Possible outcomes by letting students use technology collaboratively were analyzed in the book edited by Roberts (2003).

In the area of developing lab-based online teaching materials, some studies are for various technology based courses. Brown and Lu (2001) point out that the creation of online teaching material requires the designer to follow the instructional design theory. They developed a Web-based tutorial based on the student-centered learning theory to teach database students. Clancy, Titterton, Ryan, Slotta, and Linn (2003) describe a lab-based course format they developed, based on the research of learning science. The lab-based course format is for teaching computer science courses; it allows the instructor and students to collaborate on Web-based activities. Nokelainen, Miettinen, Tirri, and Kurhila (2002) describe a tool used to design and implement real-time online collaboration. They argue that the tool can be applied to a wide range of computer-based learning environments.

In this chapter, we are going to focus on the issues related to lab-based online teaching materials for technology-based courses such as network management and system administration.

Design of Lab-Based Teaching Materials

Like many other online courses, technology-based online courses also need to post lecture notes and other teaching materials on a Web site. In addition to the online lecture notes, technology-based online courses require more effort in developing teaching materials. These courses require lab-based teaching materials such as:

- Hands-on demonstration materials,
- Lab activity manuals,
- Online help,
- Collaborative operations, and
- Instruction for enforcing security measures on both server and client sides.

Based on the nature of the courses, the lab teaching materials can be in many different formats such as text only, text and figures combined, and multimedia based teaching materials. Each format has its own strengths and weaknesses.

Instructional design theories provide guidelines for the online teaching material development process. To make sure that the online teaching materials meet the requirements of technology-based courses, an instructor who may be the online teaching material designer and developer should complete the tasks in each phase of the ADDIE model which includes the analysis, design, development, implementation, and evaluation phases.

In the analysis phase of the ADDIE model, the instructor needs to identify the students' knowledge, learning behavior, and learning objectives. It is also important to analyze the learning environment and the timeline for completion. The design phase covers the tasks such as developing a strategy to meet the learning objectives, specifying teaching methods, selecting multimedia tools, defining the graphical user interface, and preparing online teaching materials including the course content, lab manual, syllabus, grading policy, assignments, and evaluation instrument. During the development phase, the instructor creates and assembles the online teaching materials by following the strategies and specifications set up in the design phase. The instructor may also need to develop multimedia course content and write code for GUI controls. In the implementation phase, a training program should be established to help the students remotely access the servers and to acquire the online teaching materials. The training procedure should be done at the beginning of each semester. The last phase in the ADDIE model is the evaluation phase, which includes testing and evaluating the online teaching materials. The feedback obtained in this phase can be used for further improvement.

In the following, we will first discuss the issues related to the development of lab-based teaching materials and issues related to the multimedia course content. Later, in the examples of the case study, we will apply the ADDIE model to create lab-based teaching materials.

Development of Hands-On Demonstration Materials

Hands-on practice is an important component of a technology-based course. It is necessary to provide some demonstrations for students before they can start their hands-on practice using the online computer lab. Hands-on demonstrations can be done in two ways, synchronous and asynchronous. Descriptions of these two online demonstration methods are given below.

- **Synchronous demonstration:** A synchronous demonstration is a real-time or live demonstration. All the students are required to attend the online demonstration session. During the demonstration, the instructor's computer screen is captured and posted online. The instructor gives live explanations on the hands-on activities and asks students for feedback. On the other hand, students can ask questions in real-time to clear up the ambiguous steps in the hands-on practice. There are some advantages and disadvantages about this method. The greatest advantage is that the students can directly talk to their instructor. They can get answers for their questions from the instructor immediately. Based on the feedback from the students, the instructor can decide if the students understand the course content. The instructor can slow down the demonstration process if some of the students have difficulty understanding the content. The disadvantage is that the students have to attend the presentation at a specific time.

- **Asynchronous demonstration:** An asynchronous demonstration can be viewed by students whenever and wherever they want. This is what a true online course is supposed to be. Also, the students can view the demonstration as many times as they want. However, it is not a real live demonstration. A student cannot directly have a live conversation with the instructor or with other students. Without immediate live feedback from the students, the instructor cannot adjust the progress of the presentation.

A combination of synchronous and asynchronous demonstrations is a better solution. During a synchronous presentation, record the presentation and make it available to those students who cannot make it to the live presentation. Even though these students cannot directly communicate with the instructor in a live demonstration, they can still watch the recorded demonstration at a time that is convenient to them.

Selection of Formats for Hands-On Demonstration

A hands-on demonstration can be created in different formats. The best presentation format for hands-on demonstrations is to use multimedia. Many of the Web conferencing software packages such as Macromedia Breeze and Microsoft Live Meeting can be used to carry out multimedia demonstrations.

Usually, a Web conferencing software package needs a support team for daily management and training. It also requires more computing resources. Some of the small campuses may not be able to afford it. The cost of software may also be too high for a small campus. For example, the cost of such software can be about 60K for 25 concurrent users. This is just the initial set up. Most of the Web conferencing software vendors also charge an annual fee. In such a case, a small campus can either join a nearby large campus by paying part of the cost or choose to use the format that combines text and figures instead of the multimedia format. To create a demonstration for hands-on practice, the instructor first goes through each step in a project for demonstration. Then, he/she uses screen capture software to capture the screens of all crucial steps. Next, the instructor adds the text that explains the purpose of the project and teaches how to carry out each step. After the hands-on demonstration is created, the instructor uploads it to the computer that hosts a Web server, or a file transfer protocol (FTP) server, or a learning management system (LMS) server which provides a Web site for students to download the file.

Some of the multimedia software packages allow the user to add voice and animation clips to a file for demonstration. For example, by using Breeze Presenter, one can add the voice effect to a PowerPoint presentation. This is a solution for some small campuses that are not able to support Web conferencing. If properly used, this type of demonstration can have the same effect as asynchronous Web conferencing, which is good news for those small campuses.

Demonstration content. A hands-on demonstration can cover many topics used to help students get a quick start. For each different technology-based course, the content of hands-on practice may vary. The following are some sample lists of topics for courses such as networking, database system development, application development, and Java programming.

For a networking course, a hands-on practice demonstration may include the topics such as:

- Remote access to servers dedicated for each individual student.
- Installation and configuration of network equipment.
- Configuration of network services and network applications.
- Enforcement of network security measures.

- Implementation of directory services.

For hands-on practice in a database system development course, the following topics can be included in the demonstration:

- Remote access to a server computer for each individual student.
- Installation and configuration of database servers.
- Creating database objects.
- Creating database programming units.
- Configuration of database services.
- Performing database management tasks.
- Querying information from databases.

An application development course has its own hands-on demonstration requirements. Some possible topics are listed in the following:

- Remote access to a server and client computer assigned for each individual student.
- Construction of client-server architecture.
- Development of front-end applications.
- Development of application servers.
- Configuration of database access.
- Configuration of Web services.
- Application of XML and .NET technologies.
- Development of data analysis projects.

Programming related courses are usually offered in a computer science or information systems curriculum. It is relatively easy to implement this type of course online. A hands-on demonstration of a programming related course may include the following topics:

- Remote access to a server that provides a programming software package.
- Use of programming software package.
- Executing and debugging code.

- GUI design and implementation.
- Submitting computation results for grading.

There are many other courses in a computer science or information systems curriculum that require hands-on demonstrations. Similar to the above, readers can create a list of topics to be covered in a hands-on practice demonstration based on their own needs.

Development of Lab Manuals

Today's IT industry expects our students to graduate with hands-on skills. Especially, for small companies, there is no time and money to provide training for new employees. The lab is a necessary component for most of the technology-based courses to help students gain hands-on skills. However, for an online class, the lab presents a challenge to both teaching and learning. Without face-to-face communication, it is very difficult for a beginner to find out what has gone wrong in a lab process. It is also very hard for an instructor to understand the situation on the student side. Remotely operating a computer through the Internet may add another layer of difficulty. With the difficulties in mind, the instructor needs to carefully write a lab manual with all the necessary details. The instructor can create a lab manual in various formats.

Selection of Formats for Lab Manuals

A lab manual can be created in the text format, the combination of text and figures, and the multimedia format. The combination of text and figures is a commonly used format for a lab manual. It combines the text to explain the activities and the screenshots to illustrate the key steps. The lab manual should provide clearly stated step-by-step instruction. In the text, you should emphasize each object name clicked by a mouse pointer with the bold face. Use different fonts to distinguish the code from the regular text. It also helps if a command is typed and placed in a single line by itself. You can use screenshots to show how to enter the required information into the controls such as a text box and a combo box on a graphical interface. Screenshots are useful in illustrating how to select an item using a check box or a radio button. Screenshots can also be used to show how to select information from a list box and combo box.

A lab manual can also be in the text-only format. In this case, the text should be written with different fonts and styles as mentioned in the previous paragraph. Many of us have the experience that, when we read the text-based instruction, there are always some points that are not very clear. We can interpret the text content in many different ways. A trial and error process takes a lot of time before we can figure out how to do something right. Even worse, after starting in a wrong direction, before we even know it, we have come to a dead end. Several hours or even several days of work are all wasted.

The best format for a lab manual is the multimedia format which includes sound, animation, and video clips. The illustration of the mouse movement can significantly reduce the guessing work. By using sound, an instructor can add more explanation about some tricky points and emphasize certain places where students tend to make mistakes. On the other hand, it may takes more time to produce multimedia based lab manual. For a lab manual that needs to be updated frequently, the multimedia format may not be the best choice.

Lab Manual Content

A lab manual contains instruction about how to perform lab activities. In general, a lab manual should include the following content:

- **Introduction:** The first paragraph of a lab manual is the introduction which contains the objectives of the lab activities and some background information.
- **Recommended reading:** This paragraph gives a list of chapters to read and a list of books and articles which contain the background information related to the lab work.
- **Main section:** This section shows the steps of developing a lab project. Each step contains some explanations of the lab activities and commands to be entered and the results by executing the commands.
- **Assignments:** To help students to better understand the lab activities, homework and projects should be assigned to students. In this paragraph, instructors should include information on how to submit the assignments and the due dates. Review questions can be placed after each major step or simply placed at the end of the lab manual.
- **Alerts:** During hands-on practice, there are some points where one small mistake can cause significant damage. For example, when configuring a directory service, a student may accidentally disable the service; this may lock the server so that no one can log in to the network system. To prevent this type of problem, an alert should be given in the lab manual.
- **Hints:** Sometimes, to encourage students to think actively, the lab manual may include some hints for the assignments or include some Web sites that contain related information.

Lab manuals vary from course to course and they are updated each semester. After each modification in a lab manual, be sure to test the lab manual before giving them to the students. Remember that, due to the lack of face-to-face communication in an online course, a small mistake can cause a lot of chaos.

Online Help

In an online class, adequate help for hands-on practice is critical to students' success. There are many ways to help students online. To assist students in setting up their computers correctly, a document of system requirements can be posted on the class Web site. The information about the installation of hardware and software should also be posted. During a semester, instructors can post the hardware and software update information on the Web site. They can also post notices of corrections of errors in the teaching materials.

System Requirements

Before a class starts, post the system requirements online and inform the students to verify if their systems meet the requirements. The checklist may include the following:

- **Computer hardware:** The posted document should include the requirements for the computer system such as central processing unit (CPU) speed, random access memory (RAM) size, hard disk capacity, removable storage devices, USP ports, and monitor resolution. It may also include the requirements for printing and scanning devices. To help students handle multimedia teaching materials, instructions on how to set up a microphone and speaker should also be included.
- **Network equipment:** The requirements for network equipment and its configurations, such as the configurations of network interface cards, switches, and routers, should be posted online. Instructions on the configuration of network services should also be included.
- **Software:** It should include the requirements for software and the configurations of the application software packages, client-side e-mail, database client software, LMS client software, and the configuration of the application development environment.
- **Internet connection:** For the remote access to an online computer lab, the requirement for the Internet connection speed should be emphasized. A dial-up connection is not adequate for many of the collaborative activities. It should

also include the configuration requirements for the VPN connection, remote desktop access, virtual machine remote access, database remote access, and Web service remote access.

Installation Instructions

To prepare for hands-on practice for online technology based courses, students may need to install various hardware and software. A step-by-step installation document may include the following content posted online. Or, the technical support team can provide the Web links to the vendors that have the installation instructions posted their Web sites.

- Instructions on how to install hardware devices such as hard disks, optical drives, video cards, sound cards, Web cameras, printers, speakers, and so on.
- Instructions on how to install network equipment such as routers, network interface cards, and modems, and wiring a home network. Some students may want to set up a wireless network. Therefore, the instruction of setting up a wireless network should be included, too.
- Instructions on how to install software such as operating systems, application software, multimedia software, service packs, patches, and the client software required by a database, LMS, and the Web conferencing server.

Information technology is a rapidly changing field. Make sure that the installation information is updated every semester.

Collaborative Operations

To improve the quality of online teaching and learning, students and instructors can communicate through the Internet. E-mail, VoIP, chat rooms, discussion groups, virtual whiteboards, text messengers, remote desktop access, and Web conferencing are the commonly used tools used to promote collaboration. Students can use these tools to share their experiences on problem solving and use collaboration tools provided by LMS packages for group activities. These tools can also be used to provide technical support, such as using text messengers to assist in solving technical problems and remotely accessing a student's desktop to help the student with debugging.

To use collaboration more efficiently, the instruction and arrangement of collaboration should be posted on the class Web site. The text-only format will be adequate for the instruction on collaboration. The following is some of the information that can be included in the instruction.

Schedule

In the instruction on collaboration, the instructor's office hours should be included so that students can directly contact the instructor during office hours. If there is a tutoring service in the computer lab, the lab tutoring time should also be included. During the tutoring time, students can ask the lab assistants about some technical issues or ask them to fix some problems that need to be fixed by someone physically on site. For example, after a power outage, students may need someone on site to start their servers. It is a good idea to put the meeting time for a group activity in the instruction on collaboration.

Preparation for Collaboration

To efficiently use collaboration tools, instructors can post the discussion topics on online for discussion. To make sure that each group member has done some work before participating in the discussion, instructors can give some discussion related assignments for students to work on.

Collaborative Operation

Instructors and the technical support team need to post information on how to participate a collaborative session. The following are some of the instructions that may be included:

- How to log on to a collaboration service,
- How to send invitations to other participants,
- How to upload and start a presentation,
- How to start a virtual whiteboard and hand write on a tablet,
- How to give permissions for others to ask questions,
- How to raise questions and give comments,
- How to share a computer desktop screen,
- How to allow others to operate on the currently running desktop,

- How to allow others to debug the code in a computer program,
- How to start and adjust audio conversation,
- How to start and adjust a text conversation,
- How to start and adjust a video conference, and
- How to adjust a Web camera to get the best effect.

There may be some other operations depending on what collaboration tools are used in a class. A clearly written instruction on a collaborative operation will greatly reduce the anxiety caused by operating mistakes.

Instruction on Security Enforcement

Enforcing security measures is a very important part of an online computer lab's daily operation. It is the joint effort of students, instructors, and the computer service department to keep the computer lab secure. Due to the fact that each student may be the administrators of several servers in the lab, the students or instructors may easily create a security hole in the lab. In addition to the computer service people, users in the lab need to be educated about the enforcement of security measures. Instructions and warnings on security issues should be available to all lab users. Security instructions may include the following items:

- Security problem reporting procedures,
- Information about how to keep computer systems and network systems secure,
- Announcement of security alerts,
- Information about the availability of service packs and security update patches,
- Instruction on the installation of security patches and service packs,
- Instruction on how to install and configure security software,
- Instruction on how to overcome a security vulnerability,
- Instruction on how to remove a computer virus, and
- Instruction on how to clean up an infected computer system or network system.

Security related information needs to be delivered to the users as soon as it is available. Posting a message on the Internet may not be enough. Using e-mail is a better way to deliver emergency messages.

Multimedia Course Content Development

As mentioned earlier, the multimedia format can be used for hands-on demonstrations and lab activity manuals. In this paragraph, we will investigate some general procedures on creating multimedia course content. The following are some commonly used methods for creating multimedia course content:

- Capturing activities on screen,
- Developing multimedia presentations,
- Recording and editing movies, or
- Displaying hand-written content with virtual whiteboards.

More detailed descriptions on the procedures of carrying out these tasks are given in the following.

Screen Capture

Many tools are available to carry out this task. Some of these tools can be used to capture still images on screen. There are also many tools that can record all the activities that happen on a screen. After the software is installed on a computer that will be used for demonstrations, it is relatively easy to capture the activities on a screen. To capture still images of a screen, you can follow the steps below.

- Make sure that the screen content is what you want to capture.
- Start the screen capture tool.
- Specify the screen capture type. Most of the screen capture tools give users the choice of selecting the entire screen, an open window on the screen, a scrolling window, an image on the screen, a specific region on the screen, and so on.
- Use the mouse to select an object on the screen and click the mouse to capture the screen.
- Save the file to a specific type. Most of the screen capture tools allow the user to save the image file as .gif, .png, .bmp, .jpg, .wmf, and many other types of formats. Some of them even allow the user to save the image file as a .pdf file.

It will take a little more effort to record video files for screen activities. The following is a general procedure for doing that:

- First, work through the whole process that you want to capture. By being familiar with the whole process, you will save a lot of editing time.
- Launch the screen recording tool.
- Specify what to record. The screen recording tool may allow a user to record the entire screen, an application window, or a specified area on the screen.
- Configure the recording style, sound, and other options.
- Start the recording process.
- Stop the recording by clicking the end button on the screen.
- Edit the recorded screen by adding the text interpretation, captions, hints, and so on. During an editing process, the user can also remove unwanted portions of the video.
- Save the video with a specific format. Usually, the user has the choice to save the video using a format such as Java/HTML, Flash, AVI-Movie, gif, jpg, bmp, and so forth.

The detail of recording may vary from one screen recording to another. Readers can find a specific recording process from the product manual provided by the vendor. Recorded screen activities can be added to a multimedia presentation, a document for a hands-on demonstration and a lab activity manual.

Multimedia Presentations

Tools can be used to turn a PowerPoint presentation into a multimedia presentation. Some of the tools can add both video and audio to a PowerPoint presentation while others can only add audio to a PowerPoint presentation. If an instructor has a PowerPoint presentation for a regular face-to-face class, the instructor can use these tools to quickly create a multimedia presentation and post it online. The following are the major tasks that should be done to accomplish this job:

- Create a PowerPoint presentation for a technology-based course.
- Download and install the presenter software onto the computer on which the PowerPoint file is stored.
- Plug in a microphone and adjust the sound volume. If video is to be added, install a Web camera for video recording.
- Open the PowerPoint presentation in Microsoft PowerPoint.
- After the presentation is opened, launch the audio and/or video recording process.

- Start the lecture based on the presentation slide. The instructor may include images and animations in the presentation.

- Edit the audio and video files to make them ready for deployment. The instructor may remove the silent part of an audio recording or replace some words with more proper ones. For video editing, the instructor can cut and paste video clips and trim the white space at the beginning and end of the video clips.

- The edited presentation can be saved to a local folder or upload to a Web conferencing Web site. Some of the vendors provide the auto unzip option which automatically unzips the uploaded zip file and save the presentation in a folder on the computer that hosts a Web conferencing server.

After the multimedia presentation is uploaded to a Web conferencing server, students can open the presentation online. Some of the Web conferencing servers also allow the user to save the PowerPoint presentation to a text-only file so that students can still review the course content when the multimedia one is temporally unavailable.

Usually, Web conferencing software is managed by the computer service department or online instruction support team. To use the software, instructors and students need user accounts set by the Web conferencing server administrator.

Developing Movies

A movie that contains the video and audio of a hands-on process can greatly improve the quality of online teaching. The tasks to create a hands-on demonstration movie are listed below:

- Create a video file either with a digital camcorder, a Web camera, or a screen recording tool.

- Import the video file into a computer that has a movie development tool installed.

- Decide which part of the video file to be included in the movie.

- Rearrange the video clips in the order that will appear in the final version of the movie.

- If necessary, make some adjustments such as color correction, trimming, low light adjustment, and image stabilizing.

- If necessary, create some video effects such as slow motion, subtitle, and picture-in-picture.

- Add some audio effects such as the introduction of a hands-on practice process and how to perform each step in the process.

- Edit the audio effects to adjust the volume of sound, and remove and add silent sections to accompany the video.
- Preview the movie and make necessary modifications.
- Save the movie to a local hard disk and upload it to a Web conferencing server.

A movie is a great format to demonstrate a hands-on practice process. On the other hand, it takes a lot bandwidth for Internet transmission. It is better to make a movie short and make it downloadable so that students can run it on their home computers. Saved movie files can be added to other presentation files later.

Virtual Whiteboards

A virtual whiteboard can be used in lectures, tutoring sessions, and group discussions. In a traditional classroom, the instructor uses a regular whiteboard to teach classes. On the whiteboard, the instructor writes outlines, circles important content, draws graphs to explain new concepts, and uses curves and lines to connect related content. The whiteboard is a great tool used to pass on information to students. When a group of students participates in a discussion, they can use a whiteboard to put their ideas together. For an online class, a virtual whiteboard is also a great tool for presentations and discussions. The following are some tasks that can be accomplished by a virtual whiteboard.

- A presenter can circle part of his/her presentation to emphasize the importance of that part.
- A presenter can add some quick drawings on a virtual whiteboard to illustrate his/her ideas.
- A presenter can use a virtual whiteboard to write mathematic symbols and formulas.
- A presenter can use curves or lines to connect two related contents on a whiteboard.
- A presenter can make instant comments on the side of PowerPoint presentation slides.
- If allowed, the participants can accomplish the above tasks on the virtual whiteboard. This will give the participants an opportunity to express themselves by writing and drawing.

Many of the Web conferencing software packages support virtual whiteboard functions. To write on the screen, one can either use a tablet personal computer (PC) or a tablet panel which can be connected to a computer. To use a virtual whiteboard in Web conferencing software, consider the following procedure:

- Make sure that the computer used to host a presentation is either a tablet PC or a laptop/desktop computer with a tablet panel attached to it. If all the participants are allowed to write their comments, each computer used by the participants should be either a tablet PC or should have a table panel connected to it.
- The presenter remotely logs on to the Web conferencing server through the Internet.
- The presenter creates a new meeting session by specifying the meeting name and meeting schedule which will be sent to participants through e-mail.
- The presenter and all the participants log on to join the Web conference at the scheduled time. The presenter and participants enter the meeting and select the whiteboard option to share with other participants.
- When the whiteboard session starts, the presenter specifies the pen style, the pen color, and the width of the pen point.
- Now that the session has started, the presenter can write on the whiteboard with a mouse or a tablet PC pen just like an instructor writing on a regular whiteboard in a classroom.
- When configured in the collaboration mode, other participants can give feedback through a dialog window.
- The presenter can also let the participants take control of the current screen so that they can write down their opinions.
- The content on the whiteboard can be saved for future use or can be directly e-mailed to the participants.

Another way to create hand-written course content is to first create the content on a tablet PC or a computer with a tablet panel attached to it, and then upload the content to a Web conferencing server. Microsoft OneNote is commonly used software that can handle hand writing as well as other media. With OneNote, students and instructors can accomplish the following tasks:

- Record audio from lectures.
- Store all the typed or hand written notes.
- From a Web site, students can drag text, images, charts, and diagrams into OneNote on a local computer.

- Convert the hand-written content or voice to typed content.
- Transfer the notes in OneNote into PowerPoint or Word documents.
- Distribute OneNote content directly through e-mail or upload it to a Web site.

When using it with a tablet PC, OneNote can be a great tool for college students and faculty members. It can be used to do many things that are difficult for a keyboard to accomplish. It is also a great tool to help students and faculty members carry out collaborative tasks online and to organize their lecture notes and handouts.

Case Study: Developing Online Lab Teaching Materials

As mentioned earlier, there are so many ways and different multimedia tools available to create online teaching materials for hands-on practice. Choosing a specific approach or a specific tool may not be easy. To be specific, we will create some teaching materials for an online network management course. What will be presented in this case study is a special case for illustration purposes only. Each campus may have different requirements and different tools available. The decision on what to cover in the teaching materials and how to create these materials will be based on the teaching requirements and the available tools. We will use the ADDIE model as a guideline for our development process.

Requirement analysis. The first phase in a lab-based teaching material development process is the analysis phase. We will identify the hands-on practice requirements for the network management course. We will also to collect information about the students' knowledge, learning behavior, and learning environment.

In general, the hands-on practice in the network management course will require the students to learn how to manage users, computers, and network equipment. Through the hands-on practice, the students will learn how to make a network meet the requirements for integrity, availability, and confidentiality. The following are the tasks to be performed for the network management hands-on practice.

- Create, configure, and replicate a directory service for managing users and computers at the enterprise level.
- Construct network management components such as domains, subdomains, organization units, trust relationships, users, groups, group policies, computers, and so on.

- Configure access control, network auditing, domain update, and information replication.

- Perform network administration tasks such as directory service maintenance, network troubleshooting, network failure recovery, and file distribution.

- Protect the network by identifying security vulnerabilities and enforcing security measures.

- Control remote access, set group policies, and configure network authentication.

In this case study, the online computer lab on campus currently supports Microsoft Windows and Red Hat Linux operating systems. The software in the computer lab is currently supported by the Microsoft MSDNAA program so that students can download Microsoft Windows Server 2003 from the course Web site. Most of the students have the Windows XP operating system on their home computers. As a result, they are more familiar with the Windows operating systems. Also, no one in the computer service department is really an expert on Red Hat Linux. Based on these facts, Microsoft Windows Server 2003 is the platform chosen to carry out the hands-on practice. Therefore, all the teaching materials are written based on the Windows Server 2003 platform.

As mentioned in previous chapters, each student will need two or probably three servers and will need to have the system administrator's privilege on these servers. If the class allows a maximum enrollment of 20 students, it will need 60 computers just for this class. There are not enough computer systems in the online computer lab. The requirement for the system resources is low. All it needs from a computer system is to run the Windows Server 2003 operating system. Also, it is not safe to let students be the administrator of these computers. Therefore, we decided to use Microsoft Virtual Server to create three virtual machines on each of the computers dedicated to this class. Each student will have his/her own computer with three virtual machines. When a student remotely logs on to the dedicated host computer through the VPN server, he/she will perform the network management hands-on practice on three virtual machines.

In this case study, we use the network management course as an example. Depending on the content, the course could be a junior-level or senior-level course. The prerequisite for this course is one programming language course and some introductory-level knowledge in operating systems and networking. At the junior or senior level, most of the students have developed some hands-on skills in using computer equipment. They are also familiar with the grading system and other general rules. To better understand each individual student, a preclass survey may also be distributed to the students. The survey may include questions related to the students' opinions about the online courses, their hands-on skills, content knowledge, and learning environments at home and workplace.

To make sure that the lab-based teaching materials are available, the materials must be implemented at the beginning of a new semester. Most of the content development tasks should be completed in the previous semester. During the vacation time between two semesters, these course materials should be uploaded to the servers and be made available to students once the new semester starts. The instructor may need assistance from Web masters to properly organize the course content.

Design of Lab-Based Teaching Materials

To help students accomplish these hands-on practice tasks, it is necessary to create some hands-on practice demonstrations to show students how to get the job done. The following are some topics included in the hands-on practice demonstrations:

- Demonstration of remote access.
- DNS server configuration.
- Active Directory service installation.
- Creating users and groups.
- Domain development.
- Protecting the objects in a network.
- Replicating the user and computer data stored in Active Directory.
- Setting up group policies.

Most of these hands-on demonstrations will be created in the multimedia format. We will use a screen recording tool to capture screen activities. Text-based interpretation will be added to the video clips.

To demonstrate a remote access procedure, the screen activities for creating a connection to the servers in an online computer lab are recorded. The video clips related to the following topics will be included in the demonstration:

- Configure the client side VPN.
- Log on to an online computer lab through VPN.
- Log to a host computer through Remote Desktop Connection.
- Log on to the virtual machines dedicated to each student.

Correctly configuring the DNS server is crucial for the rest of the hands-on practice. A wrongly configured DNS will fail to establish the Active Directory service.

Therefore, it is necessary to demonstrate how exactly the DNS server should be configured. The following tasks should be included in the video clips:

- Installing DNS.
- Creating a primary zone.
- Creating a secondary zone.
- Creating an Active Directory Integrated zone.

Similarly, we can create video clips for other topics such as creating users and groups, enforcing network security measures, replicating Active Directory, and configuring group policies.

After students have viewed the hands-on demonstration, they can perform their lab activities by following the guideline of the lab manual provided by the instructor. Due to the lack of face-to-face help from the instructors in an online course and due to the fact that some students may have very little experience with the content covered in the lab, more details should be added to the lab manual. With detailed information, most of the students can complete the lab work by themselves and the instructor can focus on a few of the students who have difficulty completing the lab work.

A lab session will last about one to two hours. For a two-hour lab, a huge file will be generated if the file is created in the multimedia format. Therefore, the format that combines both text and figures is a better choice for lab manuals. A lab manual can also be created with the text-only format. However, instruction in this format may not be clear for some GUI related activities. In this lab manual, we will cover the following hands-on activities.

- Collect information for the configuration of Active Directory.
- Install Active Directory.
- Configure the DNS server.

At the end of the lab manual, there are assignments which include additional hands-on activities and the instruction on how to submit the completed assignments. For example:

- Installing and configuring Active Directory and DNS for the second and third servers in the group.
- Submitting some screenshots of the lab activities through WebCT.

As part of the teaching materials, the instruction on security measures should also be posted on the course Web site. Security instruction should be concise and does not have interactivity. Therefore, it is often written in the text-only format. The sample security instruction in this case study is about installing McAfee VirusScan.

The above examples are used to illustrate the content design process. In the next subsection, we will discuss some of the issues related to the development phase.

Development of Lab-Based Teaching Materials

To create video clips that can be used to demonstrate remote accessing, follow the procedure below:

- Launch the screen recording tool, select the record option or create a new movie option.

- Determine the type of capture. In our case, choose the Full Screen option.

- If necessary, specify the size of the recording area, sound, recording mode, and other configurations for the recording option.

- If needed, specify the key strokes that initiate and end a recording, and specify the video quality, working folder, and video color mode.

- Perform the actions to create a connection to the remote VPN server located at the computer lab on campus. After the connection is established, log on to the dedicated computer in the lab through Remote Desktop Connection.

- When all the activities are completed, stop the recording. You will see a series of video clips.

- The next step is to edit the video clips. You can change the text in the text boxes which is entered by the screen recording tool automatically, and delete the video clips you do not want to include.

- After the editing, you can preview the edited screen recording. If everything is alright, save the screen recording.

One of the advantages of a screen recording tool is that it allows the insertion of text boxes so that the instructor can type in the short comments and instructions. After the instructor has entered the necessary instructions, the log on demonstration is ready to be uploaded to the course Web site for students to learn how to remotely access the servers in the lab.

Similar to the remote access process demonstration, a screen recording tool can also be used to create a demonstration on how to configure the DNS server. After the activities are recorded, add the necessary explanations in the text boxes, and the video file can now be deployed to the Web conferencing server for students to view the content.

With a screen recording tool, you can also create hands-on practice demonstrations for other topics such as creating users and groups, enforcing network security measures, replicating Active Directory, and configuring group policies.

To develop the lab manual which combines both text and figures, let us consider a sample lab manual for the configuration of Active Directory:

Network Management

Lab #2 Install and Configure Active Directory
Last Modified: 9-29-06

I.   Purpose

For Chapter II, you configured the network interface card and disk partition. Our next hands-on practice is to deal with Active Directory which implements directory services for Windows networks. By using Active Directory, you can manage domains, users, computers, and other equipment on a network. You can also use Active Directory to set up group policies, deploy application services, and enforce security measures. In this lab exercise, you are going to learn how to install and configure Active Directory.

II.   Collect Information for Active Directory Configuration

In this paragraph, we will collect the information needed by the Active Directory Installation Wizard.

- **User name and password:** You will create an administrator account on your first domain controller server with the following user name and password.

User name: studentxx   Password: Studentxx

Here, xx is the code assigned by your instructor.

- **Computer name:** Suppose each student has three servers. Their names can be set as below.

Computer Name	Domain Name
SERVERxx1	studentxx1.isc4300.lab
SERVERxx2	studentxx2.isc4300.lab
SERVERxx3	studentxx3.isc4300.lab

- For the second and third domain controller servers, a student can determine his/her own administrator name.
- The DNS server has been installed as shown in the description for hands-on practice demonstrations.

---

continued on following page

III.   Install Active Directory

In this paragraph, we will use the Active Directory Installation Wizard to install Active Directory on the first domain controller.

1.   To launch the Active Directory Installation Wizard, click Start and Run. When prompted, enter dcpromo in the Open text box, and then click OK.

2.   Once the Active Directory Installation Wizard page is opened, click Next. The next two pages are for operating system compatibility and configuration of TCP/IP. Click Next twice to accept the default.

3.   On the Domain Controller Type page, select the option Domain Controller for a new domain, and then click Next.

4.   On the Create New Domain page, select the option Domain in a new forest, and then click Next.

5.   On the New Domain Name page, enter the full DNS name shown in Figure 9-1, and then click Next. Later, the full DNS name will be used as the domain name.

Figure 9-1. Specifying full DNS name

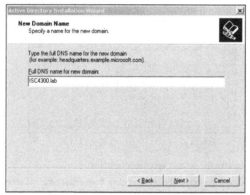

6.   When the NetBIOS page opens, type ISC4300, and then click Next.

7.   For both the Database and Log Folders page and Shared System Volume page, accept the default, and then click Next twice.

8.   On the DNS Registration Diagnostics page, choose the option Install and configure the DNS server on this computer and set this computer to use this DNS server as its preferred DNS server, and then click Next.

9.   On the Permissions page, choose the option Permissions compatible only with Windows 2000 or Windows Server 2003 operating systems, and then click Next.

10.   On the Directory Services Restore Mode Administrator Password page, enter the password, labdomain in the Restore Mode Password box, confirm it, and then click Next.

11.   On the Summary page, verify that the configurations you have made are correct, and then click Next.

13.   Click OK and restart the server.

IV.   Configure DNS Server

In this paragraph, we will configure the DNS server which the Active Directory will depend on.

continued on following page

1. To launch the DNS Manager, click Start and All Programs, Administrative Tools, DNS.

2. Right click the server name and select Configure a DNS Server from the pop-up menu.

3. On the Welcome to the Configure a DNS Server Wizard page, click Next.

4. On the Select Configuration Action page, select the option Create a forward lookup zone, and then click Next.

5. On the Primary Server Location page, select the option This server maintains the zone, and then click Next.

6. On the Zone Name page, type the zone name exactly like the one entered for the full DNS name (Figure 9-2).

Figure 9-2. Specifying zone name

7. On the Zone File page, take the default, and then click Next.

8. On the Dynamic Update page, select the option Allow both nonsecure and secure dynamic updates, and then click Next.

9. On the Forwarders page, select the option Yes, it should forward queries to DNS servers with the following IP addresses. Enter the IP address shown in Figure 9-3. Click Next.

Figure 9-3. Specifying DNS server to be forwarded to

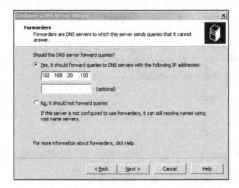

continued on following page

10. Verify the configurations and click Finish.

11. Right click Reverse Lookup Zone and select New Zone... from the pop-up menu.

12. On the Welcome to the New Zone Wizard page, click Next.

13. On the Zone Type page, select the option Primary, and then click Next.

14. On the Reverse Lookup Zone Name page, specify the network ID shown in the Figure 9-4, and then click Next.

Figure 9-4. Specifying network ID for reverse lookup zone

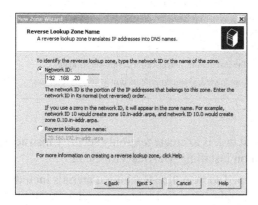

15. On the Zone File page, accept the default, and then click Next.

16. On the Dynamic Update page, select the option Allow both nonsecure and secure dynamic updates, and then click Next.

17. Verify the configurations and click Finish.

18. To add a host to the DNS server, expand the Forward Lookup Zone node, right click the zone ISC4300. lab, and select New Host(A)... from the pop-up menu.

19. In the New Host window, enter the domain name ISC4300 and a server IP address, for example, 192.168.20.1.

20. Click Add Host, and then click Done.

21. Expand Reverse Lookup Zone, right click the 192.168.20.x subnet, and choose New Pointer (PTR) from the pop-up menu.

22. In the New Resource Record window, enter the IP address 192.168.20.1.

23. Click OK to complete the configurations of the DNS server.

V.   Assignment (Due next week)

1. nstall and configure Active Directory and DNS for the second and third servers by using the information specified in this lab manual.

2. Submit two screenshots corresponding to Figure 9-2 and Figure 9-4 through WebCT.

Note: When you configure Active Directory for the second and third servers, make sure to use the existing domain ISC4300.lab instead of creating a new domain.

Box 1. Installing McAfee VirusScan

Installation of McAfee VirusScan

To protect your home computers from virus infection, you need to install antivirus software such as McAfee VirusScan and keep the software updated. The following is the instruction on how to install VirusScan.

1. Click the link to the download Web site: http://virus.uj.edu/downloads.asp

2. Download the latest version of VirusScan to your home computer.

3. If a Windows operating system is installed on your PC, remove other antivirus software, reboot your system, and then install VirusScan.

4. When prompted to enter a license option, select Perpetual licensing.

5. After you complete the installation, VirusScan will run AutoUpdate.

6. After AutoUpdate is completed, your computer is protected by VirusScan.

The next development task is to create text-only security instruction. See Box 1 for a sample instruction on installing McAfee VirusScan.

Now, we have developed some online teaching materials including hands-on practice instruction on using the multimedia format, a lab manual in the text-and-figure format, and security software installation instruction in the text-only format. Due to the size limit of the chapter, we cannot include every piece of teaching material for a network management course. In a real-life online course, much more information will be presented on the class Web site. The next phase in the ADDIE model is the implementation phase which will be discussed in Chapter X. Chapter XI will cover the evaluation phase which is the last phase in the ADDIE model.

Conclusion

This chapter has covered the issues related to creating online hands-on teaching materials. The topics such as designing online teaching materials and developing multimedia content have been discussed. This chapter has also investigated the special requirements for online hands-on practice.

To develop the teaching materials that are suitable for hands-on practice, this chapter first investigated the requirements for each type of teaching material. It listed the content to be included and the possible formats used to present these materials. This chapter has given some guidelines on how to properly match a format to a specific teaching content.

This chapter has introduced some of the methods for creating multimedia course content. It provided more detailed information on capturing activities on screen, recording and editing movies, and displaying hand-written content with tablet. Step-by-step instructions were given on how to use the tools to create multimedia materials. Lastly, this chapter presented a case study for the technology-based network management course.

After the teaching materials have been posted on the course Web site, the online class is ready to go. Our next task is to manage an online computer lab during a semester and between semesters. The related topics will be covered in the next chapter.

References

Adobe. (2006). Macromedia Breeze. Retrieved April 7, 2007, from http://www.adobe.com/products/breeze/

Bardzell, S., Bardzell, J., & Flynn, B. (2006). *Macromedia Studio 8: Training from the source.* San Diego: Macromedia Press.

Brown, J., & Lu, J. (2001). Designing better online teaching material. In *Proceedings of the Thirty-second SIGCSE Technical Symposium on Computer Science Education* (pp. 352-356). New York: ACM Press.

Clancy, M., Titterton, N., Ryan, C., Slotta, J., & Linn, M. (2003). New roles for students, instructors, and computers in a lab-based introductory programming course. In *Proceedings of the 34th SIGCSE Technical Symposium on Computer Science Education* (pp. 132-136). New York: ACM Press.

Clark, R. C., & Mayer, R. E. (2003). *E-learning and the science of instruction: Proven guidelines for consumers and designers of multimedia learning.* San Francisco: Pfeiffer.

Cole, J. (2005). *Using Moodle: Teaching with the popular open source course management system.* Sebastopol, CA: O'Reilly Media.

Koontz, F. R., Li, H., & Compora, D. P. (2006). *Designing effective online instruction: A handbook for Web-based courses.* Lanham, MD: Rowman & Littlefield Education.

Kruse, K., & Keil, J. (2000). *Technology-based training: The art and science of design, development, and delivery.* San Francisco: Pfeiffer.

Nokelainen, P., Miettinen, M., Tirri, H., & Kurhila, J. (2002). A tool for real time on-line collaboration in Web-based learning. In *Proceedings of World Conference on Educational Multimedia, Hypermedia and Telecommunications 2002* (pp. 1448-1453). Chesapeake, VA: AACE.

Rehberg, S., Ferguson, D., McQuillan, J., Eneman, S., & Stanton, L. (2004). *The ultimate WebCT handbook: A practical and pedagogical guide to WebCT 4.x.* Clarkston, GA: Ultimate Handbooks.

Roberts, T. S. (Ed.). (2003). *Online collaborative learning: Theory and practice.* Hershey, PA: Information Science Publishing.

Roberts, T. S. (Ed.). (2004). *Computer-supported collaborative learning in higher education.* Hershey, PA: Idea Group Publishing.

Southworth, H., Cakici, K., Vovides, Y., & Zvacek, S. (2006). *Blackboard for dummies.* Indianapolis, IN: Wiley.

Watkins, R. (2005). *75 e-learning activities: Making online learning interactive.* Hoboken, NJ: John Wiley & Sons.

Section IV

Management of Online
Computer Labs

Chapter X

Managing Online Computer Labs

Introduction

To keep an online computer lab fully operational is critical to the success of teaching online technology-based courses. Unlike a general purpose computer lab where heavy security measures are used to prevent users from damaging computer systems and network equipment, an online computer lab for teaching a technology-based course must release some of the security measures so that students can practice and perform an administrator's duty. That is, students and faculty members must be given the administrator's privilege to the computers and network equipment dedicated to them. For many students, this is the first time they are the administrators of network servers. You can expect that there will be many unintentional mistakes while performing the administrator's duty. Students will need help from the lab management team or the technical support personnel in the team to fix their problems and keep the hands-on process going. Some severe mistakes may cause a system to shut down without the ability to reboot. In such a case, the lab management team will need to help the students reinstall the entire system. Once upgrades to the operating system and application software are available, the lab management team will also need to install these upgrades in the lab. The content of a technology-based course will follow the change of technology on the market. Often, instructors will update the course content for each semester. These updates need to be done by the lab management team before a semester starts. Technical training is another task that will be carried out by the lab management team. Security related issues are also an important matter that should be dealt with by the lab management team. As we can see, the online computer lab maintenance tasks will keep the lab management team busy all year round. Since the security issues have been covered in a separate chapter, this chapter will focus on the maintenance tasks other than the security related tasks.

First, we will look into the issues related to daily maintenance of the online computer lab. Some general guidelines will be provided on how to manage the online computer lab and how to get help from the lab management team.

To assist the lab management, this chapter will investigate some of the lab management tools. These tools may include computer system diagnostic software, network management instruments, and so on. It will also discuss some general procedures for troubleshooting.

To prevent damage caused by a disaster such as a power outage or hurricane, this chapter will provide some information on system backup and recovery. The topics related to backup plans, system backup tools, and system backup processes will be discussed. This chapter will also provide some information on how to recover a system from a disaster.

The last topic of this chapter is about performance related issues. In general, working on a remote computer, the performance is slower than working on a local computer. It is important to maintain a computer lab in order to keep its full performance potential. The performance monitoring process and configuration tips will be discussed in this chapter.

Background

Personal computers are used in various online computer labs. The maintenance of computer hardware always deal with maintaining, upgrading, and troubleshooting personal computers (Minasi, Wempen, & Docter, 2005). The tasks in personal computer management may include memory upgrade, installation of communication devices, installation of video and sound cards, personal computer (PC) protection from viruses, capturing and editing video, power unit replacement and many other topics that are quite useful for online computer lab management. A+ certification is about the management of computer and network systems (Andrews, 2005a). Many of the skills covered by A+ certification are related to computer lab maintenance.

A big part of computer lab management is related to the management of operating systems. Operating system management tasks include planning, installing, and configuring various operating systems such as Windows Server 2003 (Snedaker, 2004) or Linux (Nemeth, Snyder, & Hein, 2002), and Mac OS (Andrews, 2005b).

The management of an online computer lab also includes the management of network systems. Network management tasks often involve network troubleshooting, network security enforcement, network update, and network redesign (Zacker & Bird, 2004). Topics related to network management such as the enforcement of security measures and maintenance of network infrastructure (Claise & Wolter, 2006) can be beneficial for the online computer labs. Network management includes the application of the SNMP packets, RMON, RMON2 MIBs, SNMPv3 standard, and the usage of device-specific enterprise MIBs (Burke, 2003). Network management tools provide great help for troubleshooting network problems. Operating systems often include some commonly used network management tools. There are also some special purpose management tools such as Ethereal for network monitoring (Orebaugh, Ramirez, & the Ethereal Development Team, 2004) and tools for managing remote access (Rasmussen & Iversen, 2005).

The management of computer and network systems is part of the general theories of IT management. The IT management theories include IT service management concepts, service management, service delivery, and security management. The information technology infrastructure library (ITIL) is a model designed for IT management professionals (Palmer, Belaire & Hernandez, 2005). Another IT management model is the Microsoft operations framework (MOF) which gives guidelines on maintaining a mission-critical system constructed with Microsoft products and technologies (Pultorak, Quagliariello, & Akker, 2003).

Computer lab management is a special case of system and network management. No matter how big or small an organization or company is. Its information system needs to be maintained by a dedicated person or team. Computer lab management requires one to have the knowledge in dealing with various technologies. There are several publications in the area of computer lab management. For example, Weeden, Scarborough, and Bills (2003) discuss some computer lab management strategies that deal with issues such as limited resources and an ever changing curriculum. Hansen (2003) reports an e-learning laboratory and management system from the banking industry that provides training in banking, insurance, and so on. In our study, we will focus on the management of online computer labs which are used to support various technology-based online courses. In the following, we will have an overview of online computer lab management.

Online Computer Lab Maintenance

Often, the computer service department will take the responsibility of managing online computer labs. The technicians in the department will perform the tasks related to technical support, training, routing maintenance, lab update, and troubleshooting. By nature, before a semester starts, there are more demands for lab update and reconstruction. During a semester, there will be some demands for troubleshooting and technical support. Upon an agreement, the computer service department will dedicate some staff members to carry out these duties. The dedicated staff members should be able to design and implement the labs to meet requirements from instructors and students. Backup staff members should also be appointed in case a dedicated person is not available due sickness or vacation.

In many cases, instructors may also be involved in computer lab management. They need to design lab activities and provide the requirements for software and hardware. They will need to work together with the lab management team to update or rebuild the labs. They will be asked to verify if the newly built labs meet the requirements of teaching and hands-on practice for the technology-based courses. They will also perform some technical support duties to help students during lab sessions.

The basic goals of lab maintenance are to:

- Make an online computer lab as secure as possible so that it is a safe computing environment to perform hands-on practice.
- Make a computer lab as reliable as possible so that instructors and students can access the lab anytime and anywhere.
- Make a computer lab able to support as many services as possible so that it can support multiple technology-based courses at the same time.

It will take great effort to achieve these goals. The following are some of the topics related to the accomplishment of these goals.

Management Model

Theoretically, there are two types of management models designed to achieve the above goals. The first management model is the IT infrastructure library (ITIL). ITIL is a library containing a set of documents and each of them covers a specific IT management topic. It provides the guideline for IT management such as incident management, problem management, change management, configuration management, and release management. The Microsoft operations framework (MOF) is another management model based on the concept of ITIL. MOF provides guidelines to ensure reliability, availability, supportability, manageability, and performance.

IT Infrastructure Library (ITIL)

For managing large-scale information technology systems, ITIL has seven topics, Service Delivery, Service Support, Planning to Implement Service Management, Information and Communications Technology (ICT) Infrastructure Management, Application Management, Security Management, and Business Perspective (ITIL, 2006). The descriptions of these topics are given below:

- **Service delivery:** Based on the agreement between the service provider and the customer, Service Delivery is a collection of management activities to make sure that a set of specified services are available to the customer. This topic contains five disciplines, service level management, capacity management, contingency planning, availability management, and financial management. Based on the Service Delivery guidelines, a set of checklists are usually prepared to help IT managers to better deliver the services.

- **Service Support:** Service Support is a collection of management activities to make sure that the users are able to access the dedicated services provided by the service providers. This topic contains six disciplines, configuration management, problem management, incident management, change management, service/help desk, and release management.

- **Planning to Implement Service Management:** Planning to Implement Service Management addresses the issues such as organizational change, cultural change, project planning, process definition, performance improvement, and where and when to start a new project.

- **ICT Infrastructure Management:** ICT Infrastructure Management provides a guideline for the activities in network service management, operations management, management of local processors, computer installation, and systems management.

- **Application Management:** Application Management is a collection of activities for improving the quality, availability, and performance of application software.

- **Security Management:** Security Management is a collection of activities to ensure the safety of information. Security measures will be enforced to preserve the confidentiality, integrity, and availability of the information.

- **Business Perspective:** Business Perspective is a collection of activities for better understanding IT service requirements. The topics covered in Business Perspective include business continuity management, surviving change, transformation of business practice through radical change, and partnership and outsourcing.

The guidelines provided by the above disciplines are the basis for developing a well-defined management process. ITIL is also the basis for other IT management models such as Microsoft Operations Framework.

Microsoft Operations Framework

This management model is for the development of a management process. It gives operation guidelines for improving the availability, reliability, supportability, and manageability of Microsoft technologies and products. MOF divides the operational activities in management processes into four quadrants, the MOF optimizing quadrant, the MOF changing quadrant, the MOF supporting quadrant, and the MOF operating quadrant (Microsoft, 2006). The operations defined in these four quadrants provide the guidelines for the management of problems, changes, plans, and assets. In each quadrant, there are three areas that are defined. The first one is the service management function which defines the management activities. The second one is the operations management review which evaluates if the management effectiveness meets the business requirements. The third area is about team roles in a management process. It describes how to form a successful team and how to relate the team work to the service management functions. The following are some further descriptions about these quadrants.

- **MOF optimizing quadrant:** This quadrant handles the issues such as how to optimize IT services to meet the business requirements. It describes how to develop service plans and how to effectively carry out the service plan. To make sure that the service management operations match the business needs, this quadrant defines eight service management functions including availability management, capacity management, financial management, infrastructure engineering, IT service community management, security management, service level management, and workforce management. In addition to these service management functions, this quadrant includes two operations management reviews, Service Level Agreement (SLA) review and change initiation review. The SLA review evaluates how effectively the management operations meet the business needs. The change initiation review evaluates the alignment of proposed changes with current IT standards and policies. Five team role clusters are included in this quadrant. They are infrastructure role cluster, security role cluster, service role cluster, partner role cluster, and support role cluster.

- **MOF changing quadrant:** To quickly and properly respond to changes in an IT process, this quadrant provides a collection of the best practices for managing changes. It also provides descriptions of management processes and responsibilities. The change management functions defined in this quadrant are change management, release management, and configuration management respectively. To minimize the impact of changes and to efficiently implement the changes, this quadrant analyzes the types of changes. Guidelines are also given on how to resolve conflicts while implementing a change. Two reviews, change initiation review and release readiness review, are also given to make sure that the changes are in alignment with the business requirements. To improve collaboration, the quadrant defines the release role cluster which includes project developers and the operations team.

- **MOF supporting quadrant:** This quadrant covers the issues related to the support of the usage of IT infrastructures. It provides the guideline for analyses and operational activities to efficiently support an on-going IT process. For supporting management, the quadrant defines the management functions such as incident management, problem management, and service desk operations. The coverage of team operations is also included. This quadrant provides guidelines for solving customer issues and team work on solving larger scale IT problems. In addition to these service management functions, one review is included in this quadrant. It is the service level agreement review. This review covers the re-evaluation of current SLAs and renegotiation with consumers when modifications to the current SLAs are necessary. For team roles, this quadrant covers two team role clusters, the support role cluster, and the partner role cluster. The support role cluster includes technical specialists and service desk personnel. The partner role cluster includes partners, vendors, outsourcing vendors, and other suppliers.

- **MOF operating quadrant:** This quadrant deals with availability and security. It contains a collection of best practices for the activities and functions in an on-going management process. The management functions defined in this quadrant are directory services administration, job scheduling, network administration, security administration, service monitoring and control, storage management, and system administration. In addition to routine management tasks such as system backup and restoration, service pack management, and security measure updates, the MOF operating quadrant also defines the operation reviews for the evaluation of operation efficiency and service optimization. There are two team role clusters, the operations role cluster and security role cluster. The operations role cluster describes the responsibilities of specialists and the security role cluster describes the responsibilities of security administrators (Microsoft, 2006).

Each of the above two models can be developed into a book. A complete description of these models is beyond the scope of this book. Readers who are interested in these two management models can go to the books and Web sites listed in the References.

Lab Usage Instructions

To achieve these goals, the first thing that the lab management team needs to do is to communicate with students and faculty members. At beginning of each semester, the team should send out lab usage instructions which may include the following content.

- Equipment usage policies and user behavior policies.
- Available hardware and software in the lab.
- Hardware and software requirements for students' home computers.
- Requirements for security measures.
- Instruction on how to report security problems.
- Instruction on how to get help from the technical support personnel and how to get help from instructors and tutors.
- Instruction on remote access.
- Course Web site log on information.
- Student e-mail account information.
- Links to course syllabi, lecture notes, hands-on practice demonstrations, lab manuals, and other teaching materials.
- Information about the usage of multimedia tools.
- Information about the usage of collaboration tools.

This information should be uploaded to the course Web site and placed on the front-page. Also, e-mail students to inform them about the availability of these instructions. Encourage students to carefully read the instructions before they start their hands-on practice and homework.

Lab Maintenance

The daily lab maintenance deals with three types of tasks, lab maintenance, lab update, and troubleshooting.

Daily Maintenance

Lab maintenance includes the tasks that are routinely scheduled to keep the lab in good condition. The following are some of the tasks:

- Check the status of the computer hardware and network equipment.
- Check the status of software.
- Check the power supply, room temperature, and other room conditions.
- Perform the scheduled system and database backup, and recover a failed system.
- Check if there is any virus infection.
- Check if there is any performance problem.
- Check if the Internet connection is working properly.
- Check if patches or updates are available for the computer hardware, software, and network equipment.

Once a problem is detected, fix the problem. If necessary, e-mail all the students about the changes made in the computer lab.

Lab Update

The technicians in the lab management team need to update the computers and network systems as soon as the security patches, software updates, and service packs are available. These tasks are not routinely performed. Another type of lab update is the update of hardware and software for next semester's teaching and hands-on practice. Once an instructor decides to update, he/she will make an appointment with the lab management team members to discuss the requirements of the new lab activities. After requirements are clearly defined, the lab management team will decide how to meet the requirements. Often, it may require adding new computers, network equipment, or new software. Within the budget limit, the computer service department will purchase the required devices or software.

After the semester is over, the technicians will update the current computer systems and network equipment. Sometimes, the technicians may need to rebuild the entire lab. For example, an instructor may decide to use a new operating system for the system administration course. By the end of each semester, the configurations of operating systems and application software may have been altered during the hands-on activities. To get ready for the new semester, the operating systems and application software have to be reinstalled. Also, between the semesters, instructors may update the course content to catch up with the current trend. It often results in rebuilding the computer lab.

During the lab update period, additional support is needed. When scheduling the workload, the manager of the computer service department should consider the update needs. If updating the computer lab requires more technicians, the manager should assign more people to accomplish the tasks.

Troubleshooting

During hands-on practice, students may wrongly configure a system and crash a hard drive or make some services unavailable. Often, instructors are more familiar with the lab activities, so they can give students some advice to fix the problem. Sometimes, they may need the technical support people to help. For example, the following situations will need help from the technicians:

- Students need help on remote access to the VPN server which is configured by the network manager.

- The hard disk on a computer used by an individual student was crashed. It needs to be re-imaged.

- Students need help to clean infected files or disks. The lab management team may have more security maintenance tools.

- Hardware and software failure may require technicians to replace the failed equipment or reinstall the software.

- Technicians are also needed to help with an Uninterruptible Power Supply (UPS) problem, local area network problem, and so on.

Troubleshooting is an open-ended task. It requires a lot of experience and careful analyses. Encourage students to provide detailed descriptions when reporting problems. If possible, ask them to provide screenshots to illustrate what has happened. Sometimes, the technicians may need to get permission from a student to remotely access the desktop of a student's computer to solve the problem.

Computer Lab Management Tools

As described above, computer lab management requires a great deal of effort. Many tools are available to handle the management of servers, networks, and security measures. Based on their functions, these tools can be classified into two categories, server management tools and network management tools.

Server Management Tools

The tools in this category can be used to accomplish the following tasks:

- **Server management:** The tools can be used for fault detection, performance tuning, antivirus protection, operating system and application software updates, directory service management, server backup and recovery, and remote software installation.

- **Server monitoring:** The tools can be used for server operation management, access auditing, event tracking, and notification service when issues arise.

Usually, most of these tools are included in a server operating system. There are also some tools designed for specific tasks and are able to be used with different operating systems. These tools are for disk imaging, virus prevention, fault detection, and for creating management reports. In the following, more information is given about the management tools.

- **Tools for disk imaging:** It may take many days to install operating systems and application software packages. Disk imaging tools are useful for reducing the installation time and for quickly restoring a failed computer system due to hard drive corruption, virus infection, and wrongly configured operating systems or applications. It can be used to image a hard disk on one computer and copy the image to the hard disks of other computers simultaneously. With a disk imaging tool, one can install operating systems and application software in a computer lab. If all the computers in the lab use Windows operating systems such as Windows XP Professional or Windows 2000 Professional, you can also use the tool Remote Installation Services (RIS) which is included in Windows Server 2003. RIS can be used to create the image of a hard disk on a source computer and copy the image to many other computers.

  There are other useful computer system management software packages that can be used for the server management in a computer lab.

- **Antivirus tool:** This tool can be used to check the files stored on disks and automatically remove spyware, viruses, Trojan horses, and worms. It can also check incoming and outgoing e-mail attachments for viruses.

- **Operating system management utility:** This type of tool can be used to recover systems from crashes, repair damaged operating systems, optimize hard drives' performance, and fix hard disk faults.

- **Activity undo tool:** This tool can be used to undo the activities performed on a computer system. It can undo the activities several days back. For example, if a student made a mistake several days ago, such as accidentally deleting an important file, he/she can still undo the wrong operation.

- **Diagnostics tool:** This kind of tool is used to check a computer system's hardware. It can also detect problems in computer system configuration and performance.

- **System optimization tool**: This tool is used to manage all operating system configurations in one place.

- **Disk management tool**: This is also a useful tool for computer lab management. It is a disk management tool. Often, in a computer lab, students may need a new hard drive partition, so they can add a new operating system to the existing system. This tool can be used to divide existing disk space into multiple partitions and each can be used to host an operating system or used to host data files or log files to improve the operating system's reliability.

Readers who want to find more information about these server management tools can go to the Web site by ServerWatch (2007). This Web site discusses server infrastructure tools, and it introduces various server management tools and the usage of these tools.

Network Management Tools

For network management, a server operating system may include some tools for network monitoring and auditing. Network management tools can be used to configure network devices and bandwidth. They can also be used to monitor network performance and detect faults in a network. The following are the tasks that can be performed by network management tools:

- **Setting up network security:** The tools can be used to enforce security measures for each tier of the network, client, midtier, server, and gateway. The tools can handle firewall configuration, intrusion detection, antivirus, and other security measures.

- **Controlling remote access:** The tools can be used to control network availability. With these tools, a network administrator can decide which hosts are allowed to access the organization's network, which protocols can run on the network, which users can access the application software, and which accessing methods are allowed.

- **Management reporting:** These tools can be used to create network management reports. A report may include the analysis of network observation data, event summaries, alerts for potential problems, and recommendations for possible solutions.

- **Network fault detection and troubleshooting:** Once a computer system problem or a network problem occurs, it needs to be fixed as soon as possible. Some tools can be used to help a network administrator diagnose network and system failures. These tools can be configured to automatically collect and analyze program and environmental failure information. To minimize the network downtime, the collected information can be used by the network administrator for faster recovery from the failure.

The tools for network security have been covered in Chapter VIII. For other tasks, some of the tools are included in server operating systems. If Cisco equipment is used for the network in the computer lab, one may find some network management tools at the company's Web site for downloading. Some commercial network management tools are also available. The following is the discussion of these tools:

- **Network monitoring tools:** For network management, there are some open source software packages which can be used for network troubleshooting, analyzing, and monitoring. They have a standard protocol analyzer that can be used for the UNIX, Linux, and Windows computing platforms. For wired and wireless network management, there are some commercial products that can also be used for protocol and trend analyses, network troubleshooting, detecting and preventing intruders, or monitoring devices and network bandwidth. In addition to performing basic management tasks such as packet capturing and analyzing, they can decode over 500 protocols and provide real-time statistics for trend analyses. Some of them can detect network problems by automatically monitoring, measuring, and controlling a network process. They can also automatically generate reports and inform the administrator of problems.

- **Network access control tools:** To protect a private network, these tools can block unauthorized devices from accessing the private network and resources. They also provide tests for current security conditions such as antivirus status, service pack versions, and personal firewall status. They can cover a wide range of computer hosts such as servers, workstations, notebooks, or PDAs, whether these hosts are managed or unmanaged by a network server.

- **Hardware diagnostic tools:** For hardware diagnosis, it is often required that the diagnostics tool is independent of the operating system. There are some commercial products can accomplish this task. These tools can perform hardware diagnosis by bypassing the operating systems such as DOS and Windows. This kind of tool is software on a bootable CD. It can identify the types of central processing unit (CPU) and BIOS. It tests the memory, audio and video devices, CD/DVD devices, network interface cards, serial and parallel ports, USB ports, hard disks, floppy disks, mice, joysticks, keyboards, modems, and so on. It also checks for temperature, fan speed, voltage reading, and PCI and plug-and-play devices.

- **Software distribution tools:** These tools can be used to distribute software through a network. They can be scheduled to automatically install software or distribute files such as scripts and service packs to hundreds of computers at the same time. They can also configure a distribution process based on the CPU performance and memory usage.

In the above, only a few types of network management tools are discussed. There are many more commercial management tools available. Many of them are free download software or allow users to evaluate for a month or longer. After testing, network administrators can decide what is the most suitable for them. For more information on network management related resources and tools, readers can refer to the Network Management Tool (2007) Web site, which provides links to the tools used in network management, Internet, and intranet.

Online Computer Lab Backup

A server operating system plays a key role for the daily operation of an online computer lab. Ideally, we would like the server to run constantly without any downtime. However, in a real-life situation, there is a chance that a disaster can happen to the server operating system. For example, a hard drive crash can make the server unavailable for lab activities. After the crashed hard drive is replaced with a new one, you need to rebuild the server. It may take days or even weeks to rebuild a server. Even worse, all the data saved on the server are lost. Since the server is shared by many users, the loss of data can have a negative impact on many users.

When a server fails because of hardware damage, software failure, or human error, it is the lab manager's responsibility to recover the system as soon as possible. To reduce the negative impact resulted from a disaster, we need to backup the server regularly. With system backups, one can recover the failed system with the least amount of time and minimal data loss. In the following, we will discuss some of the concepts and practices related to computer system backup.

In a system backup process, the currently running system can be backed up to a storage device such as a tape drive, an external hard drive, or a storage area network (SAN) device. In addition to operating systems, other services such as Web sites and databases can also be backed up so that we can easily restore these services. Some of the system backup methods are:

- Using the utilities included in an operating system. These tools are relatively easy to use and there is no extra cost.

- Using commercial backup software. Most of the backup software offers an adequate solution for network-wide backup. Some of them can also perform disaster recovery functions.

- Using a specially designed external hard disk such as the CMS external hard disk. This kind of hard disk can automatically back up a system even when a change is made to the internal hard disk. This type device is fast and easy to use, but may cost more.

- Using a hard disk image created by the disk imaging software. This method is good for initial installation of operating systems and application software on multiple computers.

- Using the backup of BIOS and the boot sector which are the two components related to the system boot process. If any of these two components are damaged, the system cannot be started. Backing up these components can at least help to get the system started.

Major IT companies such as IBM and Hewlett Packard (HP) provide backup utilities for their own computer products.

Backing up the operating system is an important task in lab management. Sometimes, it is impossible to regenerate lost data. Often, system backup is the last resort to recover a failed system. Due to its importance, two examples will be used to demonstrate how to back up a system. The first example shows a backup process with a backup tool provided by an operating system, and the second example demonstrates a disk imaging process.

Backing Up with Windows Server 2003

When the budget is limited, you can use the backup utility included in Windows Server 2003. The backup utility can handle the basic backup tasks and is simple to use. It provides the Backup Configuration Wizard for the configuration of a backup process. The following are the tasks to configure a backup process:

- Launch the backup utility.
- When the Backup Configuration Wizard starts, specify that you want to back up files and settings.
- You should also specify what to back up. You either back up the entire system or back up certain files.

- The next step is to specify the type of storage device in which to keep the backup files. You can specify a tape drive, or an external hard drive, or a shared network storage device such as SAN.

- You will need to specify the location where you can find the storage device for the backup.

- You can specify if the old backup will be overwritten by the new backup or attach the new backup to the old backup.

- You can schedule the time for the backup to take place.

- The utility allows you to specify if the backup is a normal backup which backs up all the files, a differential backup which backs up the files that have been changed since the last normal or incremental backup without clearing the archive bits, or an incremental backup which backs up the files that have been changed since last normal or incremental backup with the archive bits cleared.

- You can also configure a backup to include the data in a remote storage, to match the data before and after the backup, to compress the backup to save space, to automatically back up system states and protected files, and to disable volume shadow copies.

With the above functionalities, the free utility included in Windows Server 2003 is in fact a powerful backup tool. If students' home computers have the XP Professional installed, a similar backup utility can be used to backup the files on their home computers. In case a corrupted operating system cannot be started, one can use the tool Recovery Console to perform many tasks without starting the Windows operating system.

Creating Disk Images

Tools can be used to take an image of a hard disk and the image will be used as a backup. If the image is taken from a computer with all the software needed for a new computer lab, the image backup can be copied to all other computers in the lab. The process can significantly reduce the installation time. The following are the tasks in creating disk images:

- Verify that all the computers in the lab have the same hardware such as the same type of network interface card and the same amount of memory.

- Install the operating system and application software on one of the computers in the lab. This computer is called the source computer which will be used to create an image. Configure the software according to the requirements for the teaching and hands-on practice of a technology-based course.

- Install the disk imaging software on a lab server. Make sure that the server has enough storage space on its hard disk to keep the image.

- Launch the disk imaging software and create an image definition which specifies the name of the image and the location where the image will be stored.

- Create a boot disk which contains the DOS boot information and the network interface card driver.

- Make sure that the source computer and the server are connected by a local area network so that these two computers can properly communicate.

- On the server, create a session which will image the source hard disk in the source computer. Configure the session to accept the connection from the source computer.

- Boot the source computer with the boot disk created in the previous step. After the source computer is booted, connect to the session on the server created in the previous step.

- On the server, start the recording process to make an image of the hard disk on the source computer.

- After the image is created, configure the image session on the server to the restore mode and accept the connections from the client computers which are the computers you will copy the image to.

- Boot the client computers with the boot disks and make a connection to the session on the server.

- After all the client computers are connection to the restore session on the server, you can start the restore session.

- The restore time will depend on how large the image is and how many computers you have in the lab. Usually, after one or two hours, all the computers in lab will be installed with the new operation system and application software.

After all the computers are installed with the new software, you may need to manually change the computer name since the operating system on each of the client computers is still using the name of the source computer. To avoid manually reconfiguring the operating system, you can use Microsoft Sysprep while creating the image.

Online Computer Lab Restoration

The corruption of a hard disk or registry can make a computer fail to boot. When a computer system fails, not only does it lose all the valuable data, but also loses all the configurations of the system. If you have not made a backup for the system or there is no record of how the system is configured, it is extremely difficult to recover the original system. Another dilemma is that all the software updates, device drivers, and security patches are lost. You may have to reinstall some of them manually. The reinstallation and reconfiguration process may take several days to complete.

The process of recovering a failed system is called system recovery. System recovery is closely related to system backup. Many of the system backup software packages also include system recovery components. The following are some of the commonly used system recovery methods:

- The system recovery utility included in an operating system is an inexpensive option to instantly restore a failed system.
- Many commercial system recovery software products are also available.
- A backup device such as a CMS external hard disk can be used for system recovery. This type of device offers a solution for instant disaster recovery.
- By using a backup file, one can also restore a failed system. Although this kind of restore is not an instant restore, it can recover a failed system due to a hard drive crash or severe damage by a virus infection.

Again, large IT companies usually have their own system recovery software distributed with their computer products.

Many commercial system recovery software products and system recovery utilities can recover a failed system instantly. To be able to restore a failed system immediately, a system recovery tool must scan the system state, record the changes, and save the current system state as a restore point. Later, to recover the failed system, the system will undo the changes and return to the system state saved at the restore point. The following gives detailed information about a system recovery process.

System Recovery Point

A system recovery point is an archived system state or a snapshot. It may also be called a safe point, check point, or a system restore point, depending on the system recover tool. The system restore point is a widely used phrase which has a similar meaning to the system recovery point. However, the system restore point is only used with Windows XP and a system restore process is normally working under the condition that the operating system is still functioning. By creating a recovery point, the current state of the operating system and application software is saved to a file. During a restore process, the information stored in the restore point can be used to restore the system to the state recorded in the restore point.

There are three ways to create a restore point. You can configure a restore point to be triggered automatically by an event, to be launched at a scheduled time, or to be created manually. With a system restore tool, one can decide what events will trigger the restore tool to launch a restore point. The following are some of the events to be considered:

- A new operating system or a new application software package is installed on a computer.

- Reconfiguration of an operating system or application software.

- An operating system or application software is updated with patches, updates, service packs.

- Deletion of a file or uninstallation of a software package.

- Re-imaging a hard disk.

- Adding new hardware devices or replacing hardware devices.

Some events may not be able to create a restore point automatically. In such a case, you may need to manually create the restore point.

System recovery tools allow the user to schedule the time to automatically create restore points. Since a restore point creation process takes some system resources, users can specify the restore point creation process at the system idle time such as midnight. Users can also specify a restore point to be created every day or at different periods of time. In addition, users can specify where and how to store the created restore points.

System Monitoring

A system recovery tool allows the user to set multiple restore points for monitoring changes in a system. A system recovery tool can automatically track changes made to a computer system. It will create restore points before changes to the system occur. The restore points can be created for operating systems and application software. Most of the system recovery tools will also allow the user to manually set up restore points whenever it is needed. The following are the tasks that can be done by a system monitoring process:

- Monitor a set of selected operating system files and application software files.

- Record the file states before changes are made to these files.

- Take a full snapshot of the registry.

- Archive a snapshot of the registry of an operating system and other selected dynamic files to create restore points.

- Compress the saved system states and archived snapshots during the system idle time.

A system recovery process requires storage space to store the achieved files. Some of the system recovery tools will stop creating restore points if the storage space is below a certain level, for example, 200MB. With the system recovery tools, users can configure the maximum storage space for the achieved files. Some of the system recovery tools have difficulty storing the archived files in an external hard disk. In such a case, careful planning on the partitions is necessary to make sure that the system recovery process has enough storage space. Sometimes, you may need to use disk partition tools to reconfigure the hard drives.

System Recovery Process

Once a system comes to a halt due to errors, the most recent system recovery point can be used to recover the system to the previous state. To recover a system, one needs to perform the following steps:

- The first step is to undo changes made to the monitored files in the system monitoring process to recapture the state at the time when a system recovery point was created.

- The second step is to reconstruct the registry with the one saved in the system recovery point.

- If necessary, you may need to replace the hardware devices, and then restore the system.

- After the system is recovered, test the recovered system and enforce the security measures.

- The last step is to reconfigure the backup and the system recovery schedule, and the location of the archived files.

When the system recovery process is completed, the lab manager may send a note to everyone who has been affected by the system crash and give some prevention tips to the users so that they can prevent the same problem from happening again.

To illustrate system recovery, let us consider two system recovery processes. The first one is for Window Server 2003, and the second one is for a Linux system.

Automated System Recovery (ASR)

ASR is a system backup/recovery tool included in Windows Server 2003. ASR makes the rebuilding of a Windows based server easier by automatically restoring a failed server. Should a disaster occur, ASR can use the saved backup to recover the system to the normal state that the server was previously in. Suppose that a computer system cannot be booted and you have a backup file saved on an external storage device. The following is a demonstration that shows how to configure ASR for recovering a failed system:

- Backup all the information on the current computer with the Backup or Restore Wizard as shown in the demonstration of creating a backup. Another way to create a backup is to use the backup tool included in ASR which allows the user to only backup the information needed to recover a computer system.

- Create a system recovery floppy disk with the Automated System Recovery Wizard. The recovery disk contains information about the hardware and operating system configuration.

- To start a system recovery process, make sure that you have the saved backup files on a tape or external hard disk, the recovery floppy disk, and the operating system installation CD.

- Use the installation CD to get into the ASR restore mode. Insert the ASR system recovery floppy disk and start the system recovery process.

- The recovery process will first restore the disk configuration, the partition table, reformat the boot volume, and then copy the installation files.

- In the next step, the recovery process will install the basic version of the operating system. During the installation, you will be prompted to specify the path to the place where the backup file is stored.

- After the operating system installation process is completed, the failed system is restored and users can get their valuable data back.

- Save the log file which contains the details about the recovery process and keep the file in a safe place. Later, the file may be used for troubleshooting.

As you have seen, the process reformats the boot volume. This means that the data stored on the boot volume will be erased. To avoid this problem when installing a new operating system, keep the operating system on one drive, for example, the C drive, and store the application software and the user's data on a different drive. Once a computer system is set like this, with a backup file, a failed system can be recovered without losing valuable data.

Knoppix

Knoppix is an open source Linux system that boots and runs totally from a CD. You can use Knoppix for system recovery by rescuing the data from a crashed operating system. The following is a demonstration that shows how to configure and use Knoppix to rescue the data on a failed system:

- Install a storage device such as a hard disk which will be used to store the data files.
- On the troubled computer system, boot with a Knoppix CD. You can specify the language, desktop mode, mouse, monitor, and other devices. Make sure to specify the language as U.S.
- After the system is booted up, change the system to the read/write mode so that you can edit the configuration files.
- Assign a password for the root so that you can perform the operations that require root privileges.
- Mount the file system in read/write mode.
- Examine hardware devices such as hard disk partitions, the PCI bus, network interface cards, and so on.
- For a newly installed storage device such as a hard disk, format the hard disk with a Linux file system and partition the hard disk.
- Find the data files that you are going to rescue. Copy all the data files on the troubled hard disk to the new hard disk that has been formatted and partitioned in the previous step.

The above procedure shows how to get the data back from a failed system. After the data are moved to a safe place, you can either fix the problem in the operating system or simply reinstall the operating system to the troubled computer.

Online Computer Lab Performance Tuning

In an Internet based computing environment, many students will log on to the same server through the Internet at the same time. In such an environment, there could be many performance problems such as input/output device bottleneck, data sharing, server computer performance, network throughput, and Internet data transaction problems. To effectively support online technology-based courses, one of the computer lab administrator's tasks is to identify performance bottleneck and solve the problem by tuning the performance to make the operations in the computer lab more efficient. The following are some tasks that can be done in a performance tuning process.

- Collect the performance related information from students and faculty members who are using the online computer lab.
- Based on the collected information, identify the performance problem. There are some performance management tools available to assist the identification process.
- Solve the performance problem by altering options, changing the Internet service plan, adding new equipment, upgrading software and hardware, and modifying the network structure.
- Document the changes and inform the lab users about the changes made during the tuning process.

In the following, we will first investigate some potential performance problems in an online computer lab. Then, we will discuss some performance management tools. Lastly, we will discuss several ways to improve performance.

Performance Problems

One of the complaints from students is the performance of an online computer lab. In general, the performance of an online computer lab is slower than that of a face-to-face computer lab. An incorrectly configured computer lab can make the situation worse. Many factors can cause performance problems. A performance problem may be caused by a poor solution design, underpowered computer hardware and software, limited network bandwidth, improper configurations, or a combination of these factors. To see how these factors affect an online computer lab's performance, let us take a look at the descriptions of these problems in the following.

Solution Design

Solution architecture has a large impact on the performance of a computer lab. It is often difficult to meet everyone's performance requirements. The issues that need to be considered are:

- **Resource sharing:** It is important to know how and when a lab resource such as a server or a database is shared by multiple classes. Based on the information collected, the designer can make a design decision.
- **Load balance:** If a server gets a higher load than what is expected, the performance will decrease significantly.
- **Input/Output bottleneck:** When the system data, application data, and log data are all stored on one hard disk, it may cause a bottleneck when multiple processes read or write data to or from the disk.
- **Memory leak**: A poorly designed computer program causes memory leaks that make part of the memory unavailable for other computation.

Underpowered Computer Hardware And Software

Often, older software may cause performance problems. However, newer versions of software require much more hardware power. The following are some of the factors that can cause performance problems:

- **Software:** Older version software does not provide adequate computation power.
- **Memory:** A server system does not have enough memory, buffer, and cache to run new versions of software.
- **Hard disk:** The size of a hard drive is too small to create adequate virtual memory. Or, an older version hard disk runs at a lower speed.
- **Video Card:** An older video card has difficulty handling higher screen resolution.
- **VPN server:** A VPN server is underpowered so that it has a slower reaction to remote access.

Limited Network Capability

The following are some factors that can cause lower performance in network operations:

- **Router:** A router runs an older version protocol that cannot handle a high network transmission rate.
- **Switch:** A switch can handle only the 10Mbps/sec transmission rate.
- **Network interface card:** A network interface card can only handle the 10Mbps transmission rate.

Improper Configurations

Poorly configured hardware and software can also cause performance to slow down.

- **Software configuration:** Poorly configured software such as a database server may lock operations on the software.
- **Software conflicts:** When multiple software programs are running simultaneously, one software program may prevent others from using computer resources.
- **Driver:** A system using older drivers may have slower performance.

In the above, we have just named a few of the factors that may cause performance problems. Many other factors can cause poor performance. Identifying a performance problem is a not an easy job. Often, you need help from performance management tools.

Tools Used in Performance Tuning

By nature, it is difficult to identify a performance problem. An Internet based computer lab environment makes problem identification even harder. For many of the performance problems, unless the right tools are used, it is difficult to solve the problems. There are many performance management tools that can be used to help lab managers identify performance problems. Some of these tools are included in a server operating system; some of them are sold as commercial products, and some of them are free downloads. Some of the performance management tools can perform analysis and predict future problems. The tools can be grouped into four major categories, system monitoring tools, bottleneck identification tools, report generation tools, and historical analysis tools.

- **System monitoring tools:** These tools detect performance problems and report the problems to lab administrators. The tools can be used to monitor the performance of memory, CPU utilization, and other performance indicators such as response time, resource utilization, demand and contention, and queue length.

- **Bottleneck identification tools:** These tools can be used to identify where a performance problem occurs. They may give suggestions on which component of an operating system is experiencing difficulty or which application software may be the reason for the performance to slow down. In a network computing environment, these tools can be used to identify which individual host is the cause for a network to slow down.

- **Report generation tools:** These tools can perform detailed analyses and generate reports of the findings. With these tools, users can generate customized charts and plots to illustrate a system's performance. These tools can also provide statistical information about the performance, for example, the correlation between an application program and the CPU utilization.

- **Historical analysis tools:** These tools allow users to plan for future performance improvement by using the results of historical analyses. Some of the activities may depend on the occurrence of certain events, and some of the activities may be seasonal. By using these tools, the user can build a trend model, and use the model to predict future performance. The prediction can be used to adjust the configuration of memory, CPU, hard disks, and I/O equipment.

Tools like these can be found in various recourses. Operating systems such as Windows Server 2003 provides the tools to generate alerts, ad hoc monitoring, trend analyses, and so on. There are some alert warning commercial products and free download system monitoring tools. This type of software reports the warnings. Some of them can even make recommendations or automatically dial a phone.

Performance Improvement Measures

There are many ways to improve performance. We can redesign the structure of a system, alter the configuration of hardware and software, recode computer programs, and upgrade hardware and software. Brief discussions of these methods are given in the following.

Structure Improvement

The following are some of the structure changes that can improve performance significantly:

- To reduce the I/O bottleneck, place data files, log files, and system files on different drives.

- To improve reliability and performance, construct a RAID system with multiple hard drives.

- Construct client-server architecture to share the computing load among the front-end, back-end, and middle tier. For a large computer lab, a grid system can be constructed to share the computing load.

Hardware and Software Configuration and Upgrade

By upgrading hardware and software, you can also improve performance significantly. Some of the hardware and software configurations are also helpful.

- Upgrade the computer system with a faster CPU such as the 64-bit microprocessor from Intel or AMD to improve the performance of a large database.

- Upgrade the computer system with a bigger hard disk and more random access memory (RAM).

- Upgrade network with equipment and protocols that have better performance. Use faster network equipment such as network interface cards that have a gigabyte transmission rate.

- Turn the hardware acceleration on for better performance.

- Perform hard drive and registry defragmentation regularly to improve their performance.

- Choose a faster Internet connection service.

- Install the latest version of software.

- Upgrade software with the latest service packs and patches.

- Run the performance enhancement software.

- Run load balancing software to distribute the workload among multiple systems.

- Recode computer programs to avoid memory leaks.

Effective Use of Bandwidth

Even though we cannot physically increase the bandwidth of a network, we can still do something to improve the performance. The following are some tips to effectively use the bandwidth.

- Since compressed data take smaller bandwidth, you can compress the data before you transmit a large stream of data over the Internet.
- Configure the priority bandwidth usage to allow the important applications to get higher priority in using the available bandwidth.
- Configure the Web conferencing service to use lower resolution images so that they will take less bandwidth.
- Stop junk e-mail traffic.

In addition to what is listed in the above, there are other resources that can be used to improve performance. Many technology consulting companies provide performance tuning services. For a large campus, outsourcing the performance tuning work to a company is an option that is worth considering. Also, performance tuning is a task that highly depends on a person's experience. A lab managed by a group of experienced technicians often has better performance.

Conclusion

The topics related to the management of computer labs have been covered in this chapter. To meet the challenge in managing online computer labs, this chapter first introduced some management models. Two management models have been discussed. The first one is the IT Infrastructure Library (ITIL), which provides guidelines for IT management. The second one is the Microsoft Operations Framework (MOF), which is a management model based on the concept of ITIL.

To help the inexperienced students to get started, the next topic discussed in this chapter was about preparing instructions for lab regulations on how to properly use lab equipment and software, how to remotely access servers in an online computer lab, and how to enforce security measures. We discussed the issues related to daily lab maintenance, lab update, and troubleshooting. Some suggestions were given for remote troubleshooting through the Internet. Several lab maintenance tools have also been discussed in this chapter. These tools can be used for managing servers and network equipment.

This chapter has discussed the issues related to system backup. It has introduced various ways to make a backup. This chapter has also discussed several ways to recover a failed system. Two examples have been given to illustrate the procedure of system recovery. The first example shows how to use the built-in tool ASR to recover a Windows system. The second example is about using Knoppix to rescue the data from a failed Linux system.

The last topic covered by this chapter was about performance tuning. We have discussed some factors that may slow down performance. We also talked about some performance tuning tools and some measures to resolve performance problems.

After the lab management issues have been discussed, our next task is to evaluate an online computer lab to see if the lab meets the requirements of the online technology-based courses. The evaluation of online computer labs will be covered in the next chapter.

References

Andrews, J. (2005a). *A+ guide to hardware: Managing, maintaining and troubleshooting* (3rd ed.). Boston: Course Technology.

Andrews, J. (2005b). *A+ guide to software: Managing, maintaining, and troubleshooting* (3rd ed.). Boston: Course Technology.

Burke, J. R. (2003). *Network management: Concepts and practice: A hands-on approach.* Upper Saddle River, NJ: Prentice Hall.

Claise, B., & Wolter, R. (2006). *Network management: Accounting and performance strategies.* Indianapolis, IN: Cisco Press.

Hansen, P. (2003). The e-learning laboratory and management system FOKUS. In G. Richards (Ed.), *Proceedings of the World Conference on E-Learning in Corporate, Government, Healthcare, and Higher Education 2003* (pp. 515-518). Chesapeake, VA: AACE.

ITIL (2006). ITIL and IT service management. Retrieved April 9, 2007, from http://www.itil.org.uk

Microsoft (2006). Microsoft Operations Framework (MOF). Retrieved April 9, 2007, from http://www.microsoft.com/mof

Minasi, M., Wempen, F., & Docter, Q. (2005). *The complete PC upgrade and maintenance guide* (16th ed.). Alameda, CA: Sybex.

Nemeth, E., Snyder, G., & Hein, T. R. (2002). *Linux administration handbook.* Upper Saddle River, NJ: Prentice Hall PTR.

Network Management Tool. (2007). Retrieved April 9, 2007, from http://www.networkmanagementtoolscatalog.com

Orebaugh, A. D., Ramirez, G., & Ethereal Development Team. (2004). *Ethereal packet sniffing.* Rockland, MA: Syngress Publishing.

Palmer, R., Belaire, C., & Hernandez, A. (2005). *IT service management foundations: ITIL study guide.* Dallas, TX: Gulf Stream Press.

Pultorak, D., Quagliariello, P., & Akker, R. (2003). *MOF (Microsoft Operations Framework): A pocket guide.* Los Altos, CA: Van Haren Publishing.

Rasmussen, S., & Iversen, M. (2005). *Managing Microsoft's remote installation services.* Oxford: Digital Press.

ServerWatch (2007). Hot topics on server infrastructure tools. *ServerWatch.* Retrieved April 9, 2007, from http://www.serverwatch.com/hottopics/index.php/24031

Snedaker, S. (2004). *The best damn Windows Server 2003 book period.* Rockland, MA: Syngress Publishing.

Weeden, E. M., Scarborough, G. R., & Bills, D. P. (2003). Lab management strategies for IT database curriculum. In *Proceedings of the 4th Conference on Information Technology Curriculum* (pp. 62-66). New York: ACM Press.

Zacker, C., & Bird, D. (2004). *ALS Planning and Maintaining a Microsoft Windows Server 2003 Network Infrastructure.* Redmond, WA: Microsoft Press.

Chapter XI

Testing and Evaluating Online Computer Labs

Introduction

The process of developing an online computer lab is not a linear process. It is a cycle of developing a new lab, testing, evaluating an existing lab, getting feedback from users, and improving the existing lab. An online computer lab is a complicated project, especially if it is a medium or large online computer lab. For such a complex project, it is difficult to make everything perfect the first time. Some of the implementation may not meet the requirements exactly. As we all know, there would be some mistakes. There could be some conflicts when multiple classes share the lab resources. Also, technology changes at a rapid pace. Following the change of technology, the content of technology-based courses will change with it. As a result, equipment, software, and services will be updated accordingly. These factors will lead to the redesign or update of the online computer lab. Testing and evaluation are critical steps for meeting the ever changing requirements of teaching and hands-on practice. All in all, testing and evaluation provide guidelines for improving support for technology-based courses. In this chapter, we will discuss the issues related to testing and evaluation of an existing computer lab.

We will begin with the discussion of the requirements for testing and evaluation. The requirements include those for the testing of hardware, software, network equipment, remote accessibility, and course content. For lab evaluation, we will discuss what should be included in an evaluation report and how to collect information for the evaluation process.

Some tools are available for testing an operating system or a network. This chapter gives an overview of these testing tools. It provides some information about the usage of these tools.

Next, we will investigate various ways to carry out the testing and evaluation process. The tasks performed by a testing process may include the tests of the lab structure, lab computing environment, lab reliability, lab accessibility, lab security, hardware, software, and network devices.

Letting users test newly installed hardware and software is also important for testing and evaluation. After a software package or hardware equipment is installed or upgraded, ask the instructor to perform some activities to see if it works properly and meets the requirements of teaching and hands-on practice.

Getting feedback from students is one way to obtain the evaluation about an online computer lab. Students can be asked to provide feedback about the strengths and weaknesses of the lab. The students' evaluation can be used to further improve the lab construction and can be used by instructors to improve their online-based teaching. This chapter introduces some lab evaluation instruments that can be used to collect students' opinions.

The last topic covered in this chapter is related to the measurement of effectiveness of using an online computer lab. An effectiveness measurement process includes the selection of indicators and the values associated with each indicator. Several effectiveness measurement approaches will be discussed in this chapter.

Background

Most of the network infrastructure development processes include the stages of testing and evaluation. The tasks in these stages belong to the evaluation phase of the ADDIE model in the instructional design theory. Testing and evaluation are both necessary to correct mistakes made in the development process and to adapt to the fast changing technology. As part of the system development process, the tasks in a testing process include managing remote testing, writing bug reports, and testing outsourced projects (Black, 2002).

Several studies have been done in the fields of lab testing and evaluation. One of the early reports on lab testing is written by Simpson (1990) who describes how to test an online help system with different testing methodologies and discusses a framework for visualizing usability testing of online help systems which provide information on program commands, online tutorial, as well as user's manual. Waterson, Landay, and Matthews (2002) discuss the use of wireless Internet-enabled personal digital assistants (PDAs) to conduct usability tests. They show that the remote testing technique can identify content-related usability issues with more ease. For the evaluation of a collaboration learning system, readers can refer to the paper by Tu, Yen, Corry, and Ianacone (2003). They propose an integrated online evaluation system which combines peer evaluation, self evaluation, and student evaluation of online teaching. Their study shows that the instructors and students responded positively towards this evaluation system.

Many of the testing tasks need to be done with the assistance of testing tools. Once a network is implemented, the first thing is to test if it can function properly (Sugano, 2005). For testing a network, there are many open source performance testing tools (Blum, 2003). The Internet is often the cause of the performance bottleneck of an online computer lab. Testing a Web-based system can reveal major performance problems. A Web-based system involves various technologies. Therefore, its testing often requires an integrated approach. To perform integrated testing, the testing process should be carefully defined and various templates and checklists should be developed for the testing process (Subraya, 2006).

Testing network security is one of the major testing tasks. Network security testing includes detecting and responding to network attacks for multiple platforms such as UNIX, Windows, and Novell (Whitaker & Newman, 2005). There are many open source attack and penetration testing tools that can accomplish the testing tasks (Long, Foster, Moore, Meer, Temmingh, Hurley et al., 2005).

Testing theories and their applications are one of the subjects in the research of communicating system testing (Uyar, Duale, & Fecko, 2006). The topics such as software testing, testing nondeterministic and probabilistic systems, testing for security, and testing the Internet and industrial systems are related to our online computer lab testing projects.

Another subject to be covered in this chapter is lab evaluation which includes the theory, concepts, and technologies in computer system evaluation (Fortier & Michel, 2005), and online course evaluation which performs the evaluation and implementation of e-learning modules, courses, and programs (Khan, 2005). Specifically, the evaluation of online technology-based courses will be covered in this chapter. There are some online evaluation systems that can be used to evaluate online courses (Price, Walters, & Xiao, 2006). Many online assessment instruments are also developed for the evaluation of online courses (Achtemeier, Morris, & Finnegan, 2003).

One of the significant contributions to evaluation theory is the learning and training evaluation theory proposed and refined by Kirkpatrick (1998) and later revised by Kirkpatrick and Kirkpatrick (2005) with new strategies for effectively managing change.

Testing Online Computer Labs

After a computer lab is developed, a testing process should be used to test the hardware, software, and network. The testing process will help us identify incorrect configuration, incomplete installation, security vulnerabilities, and the weakness of the computer lab architecture. Based on the requirements on reliability, stability, portability, maintainability, and usability, a testing process performs a sequence of activities to investigate the underlying system built for the online computer lab. In general, a testing process will go through the following major steps:

- **Test planning:** In this step, you need to identify the objects to be tested. The success criteria should be specified based on the requirements. You should draft a testing procedure. According to the testing procedure, you can select testing tools, form a testing team, and schedule the test so that it will not conflict with the normal operation of the computer lab.

- **Information collecting:** At this step of the testing process, you will collect the information of the system operation statistics, communication ports, software log-on, system and application events, services provided by the servers, user's contact information, and so on.

- **Data analyzing:** After the information about the online computer lab is collected, the data will be categorized and analyzed. For security related testing, the collected data will be compared with the data stored in a vulnerability database, and then conclusions will be drawn about the findings. Through the data analysis, you may also be able to identify software bugs and misconfigurations that allow attackers to access the hardware, software, or network systems in the computer lab.

- **Test cleaning:** A testing process may change the configurations of the underlying system for testing purposes. During the test, make sure that the changes are well documented. After the test, the changed configurations in the software and hardware need to be reset back to the original state.

- **Test reporting:** The testing report should include the findings of compatibility, functionality, performance, and vulnerability. For security vulnerability, the report should also include the risk assessment of the vulnerabilities identified by the testing. Lastly, the report should also provide recommendations on how to solve the problems found during the testing.

Testing is a complicated process that requires versatile knowledge and experience. It often requires good team work to get things done. To efficiently accomplish the testing tasks, you need to have a careful plan and a detailed testing design. To implement the testing design, you need various testing tools to carry out the test and to analyze the testing results. A complete coverage of testing is beyond the scope of this book. For more information on testing related topics, readers can refer to the books in the References. For the purpose of this book, we will only discuss the topics that are relevant to online computer labs. The discussions below are about the testing of software, hardware, and security measures in a computer lab. The discussions will focus on the issue of testing tasks, commonly used testing methods, and testing tools that can be used to accomplish the tasks.

Software Testing

For the testing of software, since there is no software production involved in a computer lab development, not all the detailed software testing is needed for the software installed in a computer lab. The test will focus on tasks such as compatibility, functionality, and the performance of software products.

A compatibility test verifies if an application or a Web site is compatible with various operation systems such as different versions of Windows operating systems and different versions of browsers. The test result is very important for an online class. Students need to know the compatibility information before a class starts.

A functionality test validates that an application or a Web site can correctly perform all its required functions. By inputting a wide range of normal and erroneous data to a product, one can test the functions performed by a user interface, API, database, installation process, and so on.

A performance test can test a Web site's benchmark or scalability. By simulating the normal, peak, and exceptional load conditions, a performance test can identify performance bottlenecks and provide recommendations on performance improvement. It can also provide information to lab administrators to help them decide if the lab performance is adequate to handle the teaching and hands-on practice for an online course.

In the following, we will discuss the tests for various software products such as application software, operating systems, Web sites, and remote access utilities. The discussion will focus on the testing issues related to an online computer lab.

Application Software

For application software such as database software, office productivity software, communication software, and application development software, the testing process is focused on the application software's compatibility, performance, and functionality. The test is mainly about the configuration of the software since software itself has been thoroughly tested by the manufacturer. Let us have a closer look at the testing of the software.

- **Compatibility:** Often application software is not compatible with a different platform. For example, most of the application software for the PC platform will not work on a UNIX machine. For courses in the information systems or computer science curriculum, not all of the hands-on practice can be done through Web browsers. Some hands-on practice requires the client software to communicate with special purpose servers. In such a case, test the client software for each server.

- **Performance:** Some of the application software requires more computer resources to function properly. Sometimes, a hands-on activity may require much more computing power than the minimum requirement. The performance test of the application software with various hands-on practice projects gives a guideline for the requirement of computing resources. The performance of virtual machines is a great concern for an online computer lab. Their performance often causes the bottleneck of an online computer lab.

- **Functionality:** Sometimes, due to improper configuration or bugs in the software, students may not be able to perform some functions on their home computers. Tests should be conducted to make sure that every required function works properly on the client side. A GUI tool is an interface to communicate with a special purpose server. Test the GUI tool feature by feature to make sure that the required functions are working properly. Some of the functions will not run without installing the proper version of a service pack. Inform students about the service pack.

Operating System

An online computer lab is often constructed such that students' home computers are used as client computers which interact with the servers in the computer lab. The lab developing team can decide which operating system can be installed in the lab. However, there is not much they can do about the client computers. Therefore, testing the compatibility, functionality, and performance of various operating systems is a crucial task for online teaching.

- **Compatibility testing:** When testing an operating system, your main concern is about its compatibility. There may be various operating systems running on students' home computers such as different versions of Linux, Windows, or Macintosh. Even different versions of an operating system from the same vendor may not be compatible with each other. The older version of an operating system may not be able to use the services provided by the newer version of a server operating system. It is necessary to test the operating systems to see if they can interact with the server. At least, the computer service team should inform students which operating system is not compatible with the current server.

- **Functionality testing:** Testing a server operating system includes the test of its desktop support functionalities, the support for network operations, file transfer, remote access, and various other services. The test will inspect if the services are functioning properly under various circumstances.

- **Performance testing:** When multiple users operate on the same server, the performance of the server is a great concern. The server should have the processing power that is adequate to support online technology-based courses. The performance testing of an operating system may include the testing of file systems, data archiving, support for multiple tasks, support for multiple users, audio and video editing, image file editing, and so on.

Web Site

Web site testing includes the testing of functionality, elasticity, accessibility, and performance. To be more specific, the following are some details about each category of the testing:

- **Functionality:** In this category, we need to test if a Web site can handle JavaScript, VBScript, Active X, Plug-ins, Flash, Web services, and many other Web related components. Some of the technology-based courses may have their own requirements for these components. Testing should be done to find out if the Web site meets the requirements. Testing should also be conducted on security vulnerabilities related to these components. As we know, many of the security problems are associated with these components. Other testing may include the testing of support for audio and video functionalities.

- **Elasticity:** A test in this category is mainly for testing how flexible the course's Web site is. Browsers are the interfaces used by students to interact with the servers in an online computer lab. Some application software may have specific requirements for browsers. To make sure that students can perform their hands-on practice using an online computer lab, we must test the Web site with various commonly used Web browsers and inform students about the browsers that failed the test. Another kind of test in this category is one that tests screen resolution. Some students' home computers may not be able handle high resolution displays. After conducting the test, we need to inform students of the findings about the screen resolution requirement. The requirements for colors and fonts should also be tested. If there is a minimum requirement, we should inform students before an online class starts.

- **Accessibility:** It is important to make sure that the Web site is accessible remotely. Testing should be done on-campus and off-campus. Many of the functionalities may work properly on campus but may not work off-campus. Also, they may work for one type of remote access tool but not for other types of access tools. Broken link is a commonly encountered problem for a Web site. Make sure that each link leads to where it is supposed to go before the course materials are posted on the Web site.

- **Performance:** Web site performance is a key factor for a successful online class. Since the online teaching materials involve multimedia demonstrations for teaching and hands-on practice, the performance requirements for materials are high. The number of images, and the size and resolution of the images are also factors that may affect performance. We need to test the Web site with various Internet connection services. If it takes too much time to load the course content, it can cause frustration among students. The Web site must provide an alternative way for students who have a slower Internet connection.

Remote Access

Remote access software includes tools such as terminal service, VPN software, virtual machine remote access software, and remote access utilities provided by special purpose servers. The testing tasks may include testing the configurations of VPN, firewall service, and configurations of remote access tools on the server and client sides. Except the test of VPN, the connectivity and performance tests for other remote access tools are straightforward. The test of a VPN server includes testing the connectivity, performance, and security.

- **Connectivity:** The test of VPN connectivity is to decide if qualified students are able to gain access to a VPN server. The test will also decide if students are able to access the computers in the online computer lab. After the students successfully log on to the computers that are dedicated to them, the next task is to test if they can access the Internet through the VPN server. If a hands-on exercise requires accessing other VPN servers located on different campuses, the test should be extended to those VPN servers. For the VPN clients, the test may include testing the configuration of VPN client software, the protocol used to access remote VPN servers, and so on.

- **Performance:** The performance of a VPN server is determined by many factors. The software configuration, hardware, and network bandwidth can also have an impact on the performance. Poor performance of a VPN server can cause a bottleneck of the entire lab. If there is a VPN performance problem, it has to be solved as soon as possible. Testing the performance will help us identify the source of a problem.

- **Security:** Another important test is the test of the security of a target VPN server. Since a VPN server carries sensitive information about the internal network, it may attract Internet hackers. The test needs to determine if it is possible to gain access by unauthorized personnel to the internal network built in a computer lab through this VPN. The security measures on the VPN server such as antivirus software should also be tested to make sure that they are working properly. The test may also include testing the configuration of the firewall built on the VPN server.

Hardware Testing

A hardware testing process includes the testing of USB devices, display devices, storage devices, network devices, processors, and printing devices. The testing will collect information about these hardware devices and detect potential errors. The following are some commonly tested items.

- **Motherboard:** The hardware test for motherboard may test the video chipset, memory, plug-and-play devices, CPU, caches, serial ports, parallel ports, bus architecture, keyboard controllers, PCI slots, and power unit.

- **Network devices:** This kind of test will cover devices such as network interface cards, routers, switches, PCMCIA sockets and cards, modems, IRQs, DSL devices, wireless network devices, and network cables.

- **Storage devices:** The commonly tested storage devices are hard disks, floppy drives, optical devices, SCSI devices, IDE devices, flash drives, and external hard disks.

- **Other devices:** The hardware test can also test display devices such as monitors, video cards, and Web cams; sound devices such as sound cards, speakers, and microphones; input devices such as a keyboard and mouse; peripheral connecting devices such as a USB and FireWire.

For a computer lab, hardware testing is mainly focused on the compatibility of hardware and software, system functionality, and hardware performance. Like the testing of software, hardware testing also needs to have a test design before the testing can be started. In the test design, you will need to specify a list of specific testing criteria, the technology to be used for testing, and the procedure to get the job done:

- **Compatibility of hardware and software:** After software is installed on a computer, it may not be compatible with the some of the computer's hardware. This often happens when an older version of the software is installed on a newer computer. Some of the drivers for the new hardware devices are missing. Sometimes, some of the hardware devices are not compatible with the operating system or other application software.

- **System functionality:** A system functionality test is used to ensure that a computer system is fully functional. The system functionality test may include testing the services provided by the operating system, the device drivers, the system interfaces, and the interaction among the system components. During the test, a set of simulated users are used to perform various functionalities. If a flaw is detected, a report will be generated to inform the lab administrator.

- **Hardware performance:** The main testing task for hardware performance is the benchmark test. Among the performance testing methods, the benchmark test is commonly used to test a system or an individual component in the system. The benchmark test measures a system's performance and provides the information to assist a lab administrator in deciding if additional effort should be made to improve the hardware performance. The benchmark test can also be used for load balance testing. During the test, the underlying computer system will take a stress test such as a simulation of thousands of concurrent users. Eventually, the computer system will be exhausted. From the data collected during the test, we can identify the bottleneck in the system.

As an example on testing hardware and software compatibility, let us consider the hardware testing tool for upgrading Microsoft Windows XP to Microsoft Vista. To help users smoothly upgrade the operating system from Windows XP to Windows Vista, Microsoft provides the compatibility testing tool Windows Vista Upgrade Advisor (2007) which can be used to help Windows XP users find out whether their PCs are ready to run Windows Vista. Windows Vista Upgrade Advisor is a small program which can be downloaded from the Microsoft Web site and it requires the user to have the administrator's privilege to run the program. After the testing program is started, it will check if the software such as .NET 2.0 and MSXML6 is installed on a PC. If not, the tool will provide links to the download Web sites. The tool also checks if the PC meets the hardware and remote access requirements to run the tool. If these requirements are met, the tool will check the hardware and software on the PC. To make sure that the new operating system can handle all the peripheral hardware equipment such as USB devices, printers, external hard drives, or scanners, the user should plug in these devices before the test. The testing result will be displayed on the browser. Based on the hardware and software installed on the PC, the report will provide the recommendation on which version of Windows Vista should be installed. It also provides a list of hardware and software that do not meet the requirements to run Windows Vista. Based on the test result, the user can decide how to upgrade the PC for the next generation Windows operating system.

Security Vulnerability Testing

Since security is one of the highest concerns related to an online computer lab, after security measures are enforced, we can give them a test to see if there are security problems. Most of the security vulnerabilities are related to software. Security testing is mainly related to the issues applied to operating systems, application software, Web sites, and remote access utilities. Of these software programs, the testing will check if the following security measures meet the security requirements.

- Check if the user names, passwords, and domain names meet the requirements defined by a security policy.

- Verify if a user can log on to a server with the account that is specified by predefined role.

- After a user logs on to a server, confirm that the user can act with the privilege that is granted to him/her.

- When multiple users log on to a server, the server will create a session for each user. Make sure that each user can only work within the session created for him/her.

- Make sure the data integrity is preserved during a data transaction. There will be no intentional or unintentional modification of the original data.

- The software programs on the server are protected by antivirus software. Make sure that the antivirus software is properly installed and functioning properly.

A commonly used security test is the penetration test. By using a sequence of simulated attacks by malicious attackers, a penetration test evaluates the security of a computer system or network. The testing process identifies potential weaknesses and technical flaws in the software, hardware, and network. There are three ways to conduct a penetration test: black box test, gray box test, and white box test.

- **Black box test:** When a tester has no knowledge of the underlying system in a computer lab, the tester can perform a black box test. The test is often performed by outside evaluators who have no biased opinion about the underlying system or computer program. All that the tester needs to know is the range of the testing. To conduct a black box test, a set of inputs that simulate the attacks by an intruder are applied to the system to be tested. If the underlying system breaks down during the black box test, then the system has security vulnerabilities.

- **White box test:** When a tester has the all the knowledge of an underlying system such as source code, system and network configurations, the test is called a white box test. For example, a programmer knows every detail about his/her own code. If the programmer tests the computer program written by himself/herself, we have a white box test. Therefore, the white box test is often used for testing computer programs or software source code. In general, the white box test is an effective test in finding software bugs.

- **Gray box test:** Like a black box test, the gray box test also uses a set of inputs to simulate attacks. In addition, a gray box test also uses the knowledge about the source code and the configuration of a system or network.

If the test finds any compatibility problem, performance problem, functionality problem, or security problem, it will generate a report and submit it to the lab administrator immediately so that he/she can develop a solution as soon as possible.

Testing Technology

A testing process usually requires the tester to have adequate knowledge about the hardware, software, and network infrastructure. It also requires the tester to understand testing policies and testing criteria, and is able to analyze testing results. To assist the tester in accomplishing the testing tasks, various testing tools have been developed. These tools can be used to test the hardware, software, network, and security. The following are a few of them.

Hardware Testing Tools

Some of the hardware testing tools are full-featured testing tools and others are designed for a specific hardware device. Hardware testing tools are commonly used to test hardware performance and functionalities. The following are some of them.

- **System diagnostic:** System diagnostic tools can collect information about hardware devices such as a CPU, memory, operating system, and power source. Some of them can also provide a report about a diagnosis.

- **Hardware testing:** A full-featured hardware testing tool can be used to test hardware functionalities and detect flaws in various hardware devices such as a CPU, CMOS, PCI system, memory, IDE device, SCSI device, UBS device, and display device.

- **CPU stability testing:** It is a testing tool designed specifically for testing CPU satiability. It tests the CPU clock and the devices that interact with the CPU such as a hard disk, memory, and motherboard. It can even automatically handle a system crash.

- **RAM testing:** As the name indicates, this tool is designed to handle the test of RAM. It can be used to detect defective memory which is one of the causes of a system crash.

- **CD/DVD diagnostic:** A CD/DVD testing tool can be used to diagnose and repair a defective CD or DVD.

- **Internet connection testing:** This type of tool can be used to test various types of Internet connections such as dialup, ISDN, cable modem, and so forth. It can provide information about the network bandwidth potential and the portion that has been used by the current transaction.

- **Drive fitness testing:** This type of bootable DOS based utility can be used to test SCSI and IDE hard disk drives.

There are other hardware testing tools and many of them are open source software. Readers can find more information from the referenced books or from vendors' Web sites.

Software Testing Tools

To assist system administrators in testing software products, many software testing tools have been developed. These tools can be used to conduct tests of software performance, functionality, and Internet connection. The following are some of the commonly used tests:

- **Windows XP testing:** It is a GUI based performance testing tool written for the Windows XP operating system. It uses virtual users to test the performance of DBMS products and host servers. It will also analyze the test result and provide a report on the findings.

- **Load testing:** This testing tool can handle the load testing of large scale distributed software. In addition to performing automatic testing for performance over a large network, it allows testers to construct their test projects on top of the testing tool. It also provides graphical tools for data analyses of test results and reporting.

- **Functionality testing:** A functionality testing tool can perform system integration tests on the client-server architecture with the Windows operating system. Another use of this type of tool is to test Web site performance. Usually, a GUI is included in this type of tool so that a tester who does not have programming skills can still perform the test.

- **Web testing:** A comprehensive Web testing tool includes multiple testing components such as functionality testing, load and scalability testing, collaboration testing, and management testing. This type of tool can be used to test Web applications across multiple system platforms and various types of browsers. Besides performing a test, it also provides an intelligent reporting utility to allow the tester to view test results at different levels.

- **Web link testing:** This type of testing tool can be used to identify broken links on a Web page and analyze Web site accessibility. It provides reports on Web site response time and search engine optimization, and it helps Web masters to manage changes to their Web sites.

- **Full-featured testing:** Full-featured testing tools are versatile testing tools. Not only can they can test the performance and functionality of application software, they can also provide complete host information about the operating system configuration, hardware profile, system resource usage, performance, and the whole process at the time when a bug occurred. They can play back the captured process to give the tester a view about what happened before and after an error.

Network Testing Tools

A network can be tested in various simulated network environments. One can test a network with simulated network traffic which may include various network packets. Some network testing tools can also create network topologies and some specific network computing conditions for the tests. For testing network operations and network equipment, one can get help from the following network testing tools:

- **Network service testing:** A network service testing tool can be used to test network devices and network services by emulating a mixture of multiport, client-server traffic. To make the test close to a real-life situation, the emulated network traffic includes various network protocols integrated with different network services. The emulated network traffic can be used to test intrusion detection, system protection, Web and e-mail servers, VPN concentrators, SSL accelerators, load balancers, and so on.

- **Network equipment testing:** This type of testing tool can be used for automatic network equipment testing. It can also be used as a test management tool. It can be managed remotely through the Internet. By using this tool, a tester can run multiple tests simultaneously. As a management tool, it can create test logs and reports. The test reports can be sent to the tester through e-mail. It can also be connected to various testing equipment through their API sets.

- **Wireless network testing:** A wireless testing tool can be used to test 802.11 wireless networks. By sending wireless network traffic from each simulated client, the testing tool is able to test access points with its own hardware and IP addresses. The test can be downloaded from the vendor's Web site and is easy to install and configure. It does not require additional hardware equipment for testing.

Hardware and Software Compatibility Tools

In a computer teaching lab, the hardware may be gathered from different places. These hardware devices may not be compatible with the current operating system and application software. Sometimes, you may need to conduct a test about the compatibility between hardware and software. There are a few such types of tools available. In general, compatibility tests are used to test the compatibility among hardware platforms such as PCs, Linux, and UNIX workstations and servers. They may also check the compatibility between an operating system and application software or a Web site with different browsers. A compatibility test can be performed manually or can be done automatically with a testing tool.

A widely used compatibility testing tool is the Windows Hardware Compatibility Tests (HCT). Originally, this tool is used to assist hardware vendors in improving the compatibility with the Windows XP operating system. The tool provides tests to ensure that hardware devices and drivers are compatible with the Windows operating system. With HCT, a tester can also conduct testing such as stress testing and finding significant errors. HCT can be downloaded from the Windows Hardware Developer Central (2006) Web site.

Similarly, you can find hardware compatibility testing tools and services for other operating systems. Some of these tools are Web-based testing tools. They can test Web application software vs. a wide range of platforms such as Linux, Windows, Mac, and a variety of UNIX systems. They can be used to test a Web application under various browsers.

Testing of security vulnerability has the top priority. Due to its importance, security vulnerability testing tools were already discussed in Chapter VIII. Readers can find more detailed information about these tools in that chapter.

With different operating systems and browsers installed on students' home computers, it is crucial to develop a testing environment that includes an extensive range of hardware and software. The tester should also have adequate knowledge and skill to conduct the tests. Developing a sophisticated testing environment is often a project that is too big for a small campus to handle. In such a case, an online computer lab should be constructed with the most popular hardware and software, and then inform students to make changes accordingly. This approach may lose a few students who are not able to make changes. Outsourcing the testing tasks to a consulting company is another option. For some cost-sensitive small campuses, they need to find out if they can afford to pay for the consulting cost before a contract is signed.

Evaluation Instruments

The evaluation of online technology-based teaching provides a guideline for further improvement. The result of the evaluation can be used to improve teaching and learning, or used to improve the design and construction of an online computer lab. It can also be used as part of the needs analysis for next year's budget. The following are some objectives for an online evaluation:

- Gather feedback from students on the use of an online computer lab for teaching and hands-on practice.

- Identify the problems in computer systems, multimedia devices, and network devices that occurred during a semester.

- Identify the problems in collaboration and presentation software and hardware devices.

- Identify the needs for further improvement on the infrastructure of the online computer lab.

- Find out what needs to be improved in the online lecture notes, lab manual, and hands-on practice demonstrations.

- Check if the objectives of the online computer lab have been accomplished.

- Collect information on the effectiveness of the teaching methodology and teaching materials.

- Verify if the instructional content is suitable for online teaching and learning.

- Get feedback from students and faculty members about the efficiency of technical support.

To achieve these objectives, an online evaluation instrument can be used to collect information. It may include a survey or other means such as an online voting system. When developing an online evaluation instrument, you should follow the following guidelines:

- It should be designed so that the evaluation questions are closely related to the objectives of the evaluation. The evaluation instrument should check if an online computer lab meets its design goals and if students have learned as much as they would have in a face-to-face class.
- The evaluation questions should be clearly written and be displayed in a well-defined format. The questions should be classified and each group of questions should be used for a similar topic.
- The evaluation should cover every aspect of an online course. It should include the evaluation of the course, evaluation of faculty members, and the evaluation of the computer lab. It should include questions for the course content, teaching materials, and lab environment. It should cover how an online computer lab supports the hands-on practice and how the multimedia teaching materials improve the Web-based teaching.
- The questions in an evaluation instrument should follow the university's regulation. Students' privacy should be protected. The questions should be culturally sensitive.
- The evaluation results should be archived and categorized for analysis. The results should be available to the lab administrator and faculty members so that they can make modifications based on the results.

As an example, the following are some of the questions that can be included in an evaluation instrument. These questions are categorized in four areas: the evaluation of the course, evaluation of students' behavior, evaluation of instructors, and evaluation of the computer lab. To respond to the evaluation questions, one can select one of the answers from a predefined rank from strongly disagree to strongly agree.

Course Evaluation

In this area, a survey can be used to give course related statements shown in the following. Students can select one answer from multiple choices:

- The course objectives were clear to me at the beginning of the course.
- The course has met all the objectives.
- The teaching materials are valuable in helping me understand the content.
- The class Web site is well organized.
- The tools provided on the class Web site are easy to use.
- The collaboration tools are helpful.
- The class assignments are beneficial.
- Overall, I am positive towards an online course.

In addition to the multiple choice statements like the above, the evaluation may also add some questions for students to give their comments. Some of the questions related to the course are:

- What is the strength of the course?
- What is the weakness of the course?
- Does the course encourage collaboration?
- Would you like to take an online cause like this one again?
- Any constructive suggestions for making the course better?

Evaluation of Students' Behavior

In this area, statements can be given to ask students' experience in the online learning. The following are some of the statements.

- I can keep up with the pace of the class.
- I never miss any assignments.
- I am actively involved in the collaborative learning.
- When I have a problem or question, I like to contact my classmate or my instructor.
- I always carefully review the course content before taking a quiz or exam.

For questions on students' behavior, we may have the following:

- What is your strength in dealing with a technology-based online course?

- What is your difficulty in dealing with a technology-based online course?
- Do you have the prerequisite knowledge required for this course?

Evaluation of Instructors

Like a face-to-face class, the evaluation of instructors is always an important component of the course evaluation. For the evaluation of a faculty member, include:

- The instructor is knowledgeable about the course content.
- The instructor is knowledgeable about the hands-on practice.
- The instructor knows how to solve my technical problems.
- The instructor knows how to explain difficult concepts.
- The lecture notes are well prepared by the instructor.
- The instructor always returns my e-mail within 24 hours.

You can also ask students the following questions:

- Do you clearly understand the instruction given for all the assignments?
- Do you clearly understand the grading policy of the instructor?
- Do you get help from the instructor outside the office hours?
- Does the instructor grade your assignments promptly?
- Would you like to recommend the instructor to other students?

Evaluation of Computer Lab

Statements about using an online computer lab for hands-on practice are included in this part of evaluation. The following are some statements that can be used for lab evaluation.

- I can always remotely access the online computer lab.
- I can do all my hands-on assignments through the online computer lab.
- The performance of the online computer lab is adequate.
- The technical support is helpful.
- The multimedia hands-on practice demonstrations are beneficial.

- The lab manual is clearly written.
- I have learned a lot of hands-on skills by using the online computer lab.

Again, some of the lab operation related questions are given below:

- Did you get the lab operation instruction at the beginning of the semester?
- Are the technologies provided by the lab adequate to you?
- Do you clearly understand the lab security rules?

For the evaluation of an online computer lab, instructors' feedback and technical support team's opinions are as important as the students'. Usually, having regular meetings including faculty members and computer service personnel is a better way to exchange information about how the computer lab is doing and whether there are problems in the lab management. A similar online evaluation form for the online computer lab can also be created for instructors if the university has several campuses or it is difficult to organize regular meetings between faculty members and computer service personnel. E-mail is also a convenient tool for instructors to send feedback to the technical support team.

Create Online Evaluation Form

An online evaluation instrument can be constructed with an online form which contains check boxes or radio buttons for multiple choice questions and combo boxes for selecting a list of items. Text boxes are also needed for students to enter their comments about the teaching and learning with an online computer lab. A short instruction on how to fill out the evaluation should be placed at the beginning of the form. Web page development tools such as Microsoft FrontPage or Macromedia Dreamweaver can be used to create these online evaluation forms.

Deploy Online Evaluation Form

The online evaluation form should be posted on the class Web site. To store data entered by students, a database should be designed and implemented to support the form by automatically storing the data. To assist analysis, connect a spreadsheet utility to the database and use the spreadsheet's analysis tools. With the spreadsheet, one can run various statistical analyses and create pie charts, bar charts, and other plots to present the testing results. Finally, create a report to explain the findings.

Measuring Effectiveness of Using Online Computer Labs

One of the measures of an online computer lab is the effectiveness. An online computer lab takes less room space and is more scalable. It can reach more students who otherwise cannot take the class due to the inconvenience of location and schedule. On the other hand, it requires more remote access hardware and software. It needs more skilled technicians to construct and manage it. It also needs a strong technical support team to support the instructors and students. After a university puts so much effort into the online computer lab, the question is how we can make sure that it is effective. The following are some questions we can ask:

- How much have the students learned?
- Does the computer lab really do what it is expected to do?
- Overall, do students like to practice hands-on skills online?
- How many students dropped out of the online class?
- Is the cost of the online computer lab justified?
- How much extra effort does an instructor have to make to teach the online class?
- Has the enrollment increased due to the convenience of using the online computer lab?

To answer these questions, one needs to collect data related to the effectiveness of a technology-based course using an online computer lab. The following will investigate some effectiveness measurement issues.

Measurement of Effectiveness

In general, effectiveness can be measured by the cost and the usage of an online computer lab. The following are some areas where you can measure the effectiveness.

- Students' scores from exams, quizzes, and assignments.
- Ranking scores from the questions on the course evaluation survey.
- Class enrollment statistics.
- Annual cost on equipment purchasing, lab management, and technical support.

- Instructors' annual evaluation.
- The number of the courses and students supported by the online computer lab.

Similar to face-to-face classes, the measurement of effectiveness will mostly depend on the data collected from the above areas. Therefore, we need to take a closer look at these areas in the following.

- **Scores of exams, quizzes, and assignments:** Students' scores from the exams, quizzes, and assignments are an important indicator of how much students have learned from a class. To measure the overall performance, the exams, quizzes, or assignments should include some hands-on related questions. To get unbiased results, the difficulty level of the questions should be equivalent to that in a face-to-face class. One of the difficulties for giving an online exam is that there is no easy way to give a closed-book exam online. One of the solutions is to let students go to a designated location to take a proctored exam. For universities with campuses spread out in rural areas, this solution may not work well. Another solution is to set up a time limit for an open-book exam so that students have less time to find answers in the book.
- **Ranking scores of course evaluation survey:** The ranking scores in a course evaluation survey are valuable in measuring the effectiveness of an online course. These scores can be used to decide if an online computer lab has met the requirements for teaching and hands-on practice. They can also be used to see if students really like the course. Therefore, when designing a class evaluation survey, make sure to include items that are related to effectiveness.
- **Enrollment data:** Enrollment data can be obtained from a university's registration office or institution research department. The enrollment data is a good indicator of the effectiveness of courses. The enrollment data can also be used to find out how many courses are supported by a computer lab and how many students are enrolled in these courses. However, other factors can also cause the enrollment to go up and down. For example, the enrollment in technology-based courses is greatly influenced by the job market trend. When the technology job market is down, the enrollment in these courses can decrease dramatically; this has nothing to do with the quality of teaching or the effectiveness of the computer lab. Also, advertising the courses is another issue that can affect enrollment. Students may not know the existence of these online courses, or they do not know how to enroll in these courses. Sometimes, small things such as a misprinted prerequisite can significantly reduce the enrollment. Therefore, when enrollment is down, check if it is caused by these factors before making conclusion on the effectiveness of a course. On the other hand, when enrollment is up, make sure that it is not just because of the job market trend.

- **Online computer lab expense:** The online computer lab expense is another indicator to measure the effectiveness of an online computer lab. When using the expense as an indicator of effectiveness, we need to distinguish the initial cost and annual cost. The initial cost is a one-time investment and it can be much bigger than the annual cost. When calculating the cost, not only do we need to count the cost of equipment, but we also need to add the cost of service personnel. Also, do not forget the savings by using the online computer lab. Usually, we can obtain the cost information from the computer service department.

- **Evaluation of instructors:** In the faculty evaluation, you may find detailed information about how much extra effort a faculty member has put into the development of online teaching materials, sometimes including developing an online computer lab. The faculty annual evaluation may also reveal how the effort on developing an online computer lab has affected a faculty member's research and other services. For some universities, there may be some compensation for the online course development. You may also find this type of information by giving a survey.

Putting all the data from the above measurement together, we will have a whole picture on the effectiveness of an online computer lab and the online course. Once the data are collected, our next task is to analyze the data by various means.

Effectiveness Assessment Methods

The goal of analyzing effectiveness is to see how many objectives have been accomplished after a certain amount of money has been spent. However, unlike the effectiveness analysis in business, the effectiveness analysis for online computer labs does not consider the money spent on a computer lab project as purely the annual cost. The expense is considered a long-term investment. Also, it is difficult to assess the effectiveness by next year's student enrollment. There are different ways to analyze the data. It can be a descriptive analysis or a statistical analysis. In a descriptive analysis, the following methods can be used to describe the collected data.

Descriptive Methods

To explore the collected data, you can use graphic descriptive methods such as charts, plots, tables, and numerical descriptive methods such as means, medians, modes, and standard deviations:

- **Charts:** There are various charts that can be used to analyze the data. These charts are used to summarize the characteristics of a data set. To describe the data collected from a single qualitative variable such as the major of students in each class, you can use a pie chart or a bar chart. For a single quantitative variable, you can use a pie chart, bar chart, and histogram. To describe the relationship of a two-dimensional variable, you can use a scatter plot, x-y plot, and bar chart.

- **Tables:** A frequency table is a commonly used table to describe the data collected from a single or a two-dimension variable. For the data collected from two-dimension variable, you can also use cross-classification tables.

- **Numerical descriptive methods:** To describe the central tendency of a variable, you can use the methods such as the mean, median, and mode. To measure the data variation, you may consider using the range, variance, and standard deviation. When using descriptive statistics for a small class, the mean of the ranking scores may give a misleading conclusion due to a few outliers. In such a case, using the median will provide a more reliable result.

The outcome of the analysis will be used to assess the effectiveness of the online computer lab. Sometimes, more advanced statistical methods are used to analyze the relationship among variables.

Statistical Methods

Statistical methods may include the correlation of two variables, meta-analysis, and the cost-effectiveness ratio.

- **Correlation:** Correlation is used to measure the strength of a relationship. For a large sample, one can calculate the Pearson product moment coefficient of correlation. For small samples, the measurement of effectiveness can be based on a ranking system which includes a statistical analysis on the ranks entered by students on the evaluation form.

- **Meta-analysis:** Meta-analysis is a statistical method that can be used to combine the results of several similar statistic hypothesis tests in order to draw general conclusions. With meta-analysis, the result of the analysis is more accurate for a small sample size. The meta-analysis method is widely used in the area of medical research and educational research.

- **Cost-effectiveness ratio:** This ratio measures the value of project costs vs. the value of project outcomes. The lowest cost-effectiveness ratio indicates the project has the best efficiency. There is often a trade-off between effectiveness and impact. The decision depends on the emphasis of the university. It also provides the baseline for assessing how much more it would cost in terms of extra resources to achieve greater results, through the use of more effective but more costly alternatives.

The analysis will reveal the effectiveness of the computer lab. A detailed description of any of these statistical methods can occupy a whole chapter. Further discussion of these methods is beyond the scope of this book.

Kirkpatrick Four-Level Evaluation Model

Kirkpatrick's evaluation model consists of four levels: reaction, learning, behavior, and results (Kirkpatrick, 1998; Kirkpatrick & Kirkpatrick, 2005).

The first level of this model is the reaction level, which can be described as how participants feel about a training program. For the evaluation of an online computer lab project, survey questions related to how students feel about the course content, hands-on practice, and online computer lab belong to this level. If students do not have a positive attitude towards an online technology-based course, they will not put full effort to the study of the course material and they will not get adequate skills from hands-on practice.

Different from corporate training, the curriculum in a higher-education institution is predetermined and it is less flexible to adding or removing courses from the current curriculum. However, instructors can make more effort in emphasizing the importance of the courses and give better descriptions about the objectives of the hands-on practice.

The second level of this model is learning. This level of evaluation is about the measure of knowledge acquired or attitude changed. Many forms of measurement tools have been used in higher education institutions, such as pre-assessment survey, placement exams, grades, and post-assessment survey. For an online computer lab, the lab accessibility, functionality, and availability should be evaluated together with the measurement of hands-on skill improvement.

The third level of this model is behavior. This level of evaluation is about the behavior of the students. The evaluation is used to find out how students' behavior changed in following the instructions of hands-on practice materials and in using the lab equipment. It can also be used to evaluate how the activities in the lab have changed the students' behavior in the areas such as collaboration and self-discipline. This level of evaluation is to evaluate whether the students have used the new knowledge, skills, and attitudes in their learning process. The knowledge and skills should be applied to the subsequent courses and jobs.

The fourth level of this model is about results. The outcomes of this evaluation will reveal if the learning has an impact on the students' careers, professional development, and business environments. Alumni surveys and employer feedback can be used as evaluation tools.

The Kirkpatrick four-level evaluation model has been widely used in many technical training organizations. By following the Kirkpatrick four-level evaluation model, one can design a comprehensive and efficient evaluation process for an online computer lab. There are several other evaluation models. Further discussion of evaluation theory is beyond the scope of this book.

Conclusion

This chapter has covered the topics related to lab testing and evaluation. First, an online computer lab must be tested to verify if the lab meets the requirements of a technology-based course. Based on their functionalities, tools for testing were also reviewed in this chapter. Topics on testing the lab structure, lab computing environment, lab reliability, lab accessibility, lab security, hardware, software, and networks were also covered in this chapter.

The next topic covered in this chapter is evaluation which is a process to collect users' opinions on the online computer lab. We have discussed the design of an evaluation survey which includes performance related statements and questions. This chapter has introduced several ways to collect evaluation data.

The last topic covered in this chapter is about measuring the effectiveness of the online computer lab. Some descriptive methods and statistical methods were mentioned in this chapter.

At this point, we have discussed the stages of online computer lab development and management. The topics related to the design, implementation, maintenance, security, course content creation, testing, and evaluation have been discussed throughout the book. From the instructional design theory point of view, we have accomplished the tasks in all the phases of the ADDIE model. The final topic discusses the trends of online computer labs, which will be covered in the next chapter.

References

Achtemeier, S. D., Morris, L. V., & Finnegan, C. L. (2003). Considerations for developing evaluations of online courses. *Journal of Asynchronous Learning*

Networks, 7(1). Retrieved April 9, 2007, from http://www.aln.org/publications/jaln/v7n1/v7n1_achtemeier.asp

Black, R. (2002). *Managing the testing process: Practical tools and techniques for managing hardware and software testing* (2nd ed.). New York: Wiley.

Blum, R. (2003). *Network performance toolkit: Using open source testing tools.* New York: Wiley.

Fortier, P. J., & Michel, H. E. (2005). *Computer systems performance evaluation and prediction.* Burlington, MA: Digital Press.

Khan, B. H. (2005). *Managing e-learning strategies: Design, delivery, implementation and evaluation.* Hershey, PA: Information Science Publishing.

Kirkpatrick, D. L. (1998). *Evaluating training programs: The four levels* (2nd ed.). San Francisco: Berrett-Koehler Publishers.

Kirkpatrick, D. L., & Kirkpatrick, J. D. (2005). *Evaluating training programs: The four levels* (3rd ed.). San Francisco: Berrett-Koehler Publishers.

Long, J., Foster, J. C., Moore, H. D., Meer, H., Temmingh, R., Hurley, C., Bayles, A. W., Liu, V., Petruzzi, M., Rathaus, N., & Wolfgang, M. (2005). *Penetration tester's open source toolkit.* Sebastopol, CA: Syngress.

Price, T., Walters, J., & Xiao, Y. (2006). A role-based online evaluation system. Retrieved April 9, 2007, from *http://www.iadis.org/Multi2006/Papers/15/F023_EL.pdf*

Simpson, M. (1990). How usability testing can aid the development of online documentation. In *Proceedings of the 8th Annual International Conference on Systems Documentation* (pp. 41-48).

Subraya, B. M. (2006). *Integrated approach to Web performance testing: A practitioner's guide.* Hershey, PA: IRM Press.

Sugano, A. (2005). *The real world network troubleshooting manual: Tools, techniques, and scenarios* (Administrator's Advantage Series). Hingham, MA: Charles River Media.

Tu, C., Yen, C., Corry, M., & Ianacone, R. (2003). Collaborative evaluation of teaching for e-learning. In G. Richards (Ed.), *Proceedings of World Conference on E-Learning in Corporate, Government, Healthcare, and Higher Education 2003* (pp. 325-330). Chesapeake, VA: AACE.

Uyar, M. Ü., Duale, A. Y., & Fecko, M. A. (Eds.). (2006, May 16-18). Testing of communicating systems. In *Proceedings of the 18th IFIP TC 6/WG 6.1 International Conference, TestCom 2006,* New York, New York. New York: Springer.

Waterson, S., Landay, J. A., & Matthews, T. (2002). In the lab and out in the wild: Remote Web usability testing for mobile devices. In *CHI '02 Extended Abstracts on Human Factors in Computing Systems* (pp. 796-797).

Whitaker, A., & Newman, D. (2005). *Penetration testing and network defense.* Indianapolis, IN: Cisco Press.

Windows Hardware Developer Central. (2006). HCT 12.1 for "designed for Windows" testing. Retrieved April 9, 2007, from http://www.microsoft.com/whdc/devtools/HCT12.mspx

Windows Vista Upgrade Advisor. (2007). Make sure your computer is ready for the edition of Windows Vista you want. Retrieved April 9, 2007, from http://www.microsoft.com/windows/products/windowsvista/buyorupgrade/upgradeadvisor.mspx

Section V

Trends and Advances

Chapter XII

Trends in Online Teaching of Technology-Based Courses

Introduction

Online (or Web-based) teaching has become one of the main approaches to delivering knowledge. As mentioned in Chapter I, almost all of the universities have adopted online teaching. Certainly, we have entered a new era for higher education. As the momentum builds up, more and more technology-based courses are also taught through online classes. Online computer labs are a key element to support the server-side hands-on practice in an online technology-based class. Due to the fact that an online computer lab is highly related to the advance of technology, lab-based online teaching still has a lot of areas that need to be improved. With that in mind, we will investigate the trends in Web-based teaching and in computer lab related technologies in this chapter.

First, we will discuss the trends in Web-based teaching, which provide a framework about the future development in this area. The topics such as future e-learning structure, management, and content development will be discussed.

Next, the trends in lab development technology will be explored. Technology is a rapidly changing field. Technology-based courses often require computer labs to following these changes. When developing an online computer lab, we must carefully examine possible changes in technology. In this chapter, we will investigate the trends in the software, hardware, and network technologies, which are potentially useful for the development of an online computer lab.

The trends in technology have a direct influence on the development of online computer labs. Not only will an online computer lab provide the training for using new technologies, it will also use the new technologies to make itself more reliable, more accessible, and faster. In this chapter, we will discuss some potential uses of the new technologies in an online computer lab.

Background

Web-based teaching will continue to be one of the major teaching methods in higher education and will keep growing. There are a number of surveys and research articles that analyze the trends in Web-based teaching. The survey report by the Sloan Consortium (2005) indicates that there is a strong upward trend among universities to consider online education as a long-term strategy. Most of the universities consider online education to be critical for their institutions. Paulsen's (2005) presentation discusses the trends in e-learning. He predicts that globalization and large-scale operation are the trends for online education. He also provides some recommendations for institutional strategies, pedagogy, cost effectiveness, and sustainability. Reiser and Dempsey (2006) discuss the trends in instructional technology and instructional design. Readers can find more information about the past and current trends in instructional technology and design from their book. They also discuss the issues that will possibly impact the future.

The report by the New Media Consortium and EDUCAUSE Learning Initiative (2006) points out that six emerging technologies will expand the boundaries of Web-based teaching and learning. These technologies are social computing, personal broadcasting, mobile phones, educational gaming, augmented reality, and context-aware environments and devices.

Ray (2006) provides information about short-term trends in technology. He points out that the technologies in voice over the Internet, mobile computing, hosted applications, data backup, security, and so on will continue to be the trends.

Research on the integration and presentation of data from diverse sources on the Web will also have an impact on the future Web-based teaching. The Semantic Web is a subject that deals with this issue. It allows browsers (or software agents) to find, share, and combine information from diverse resources, which is a great feature that can make Web-based teaching more flexible and efficient. Koper (2004) explores the application of the Semantic Web in education. He points out that the Semantic Web allows teachers to work under a flexible online teaching environment. It also allows a software agent to construct a distributed, self-organized, and self-directed lifelong learning environment. Camacho and R-Moreno (2007) proposed a new metadata-based model for learning design. This model can implement an annotated course by incorporating the course content with up-to-date semantic information. It can automatically assign instructors and schedule student activities. Based on the results of learning, it can also automatically modify the learning design to improve learning. Ullrich (2005) also discusses a course generator which can automatically assemble the learning objects from different sources to create a new course "based on pedagogical tasks and methods, formalized in a hierarchical task network (HTN) planner" (p. 74).

Trends in Web-Based Teaching

In the next few years, Web-based teaching will continue to be one of the main methods in higher education. Once students can perform their hands-on practice online, more and more technology courses will be taught online. Some of the technology-based courses can even be offered globally. Learning management systems will continue to be improved and better collaborative learning environments will be created. There will be some new ways to develop course content so that the teaching materials will be more suitable for online courses. The following are some further discussion in these areas.

Globalization

Technically, it is not very difficult to globalize an online course. Through the Internet, students can access the course Web site from anywhere in the world. However, there are many management problems that need to be solved. Since students will register for classes around the world, a university that is offering them has to deal with the differences in the following subjects:

- **Cultural difference:** Students are from different regions and they live under different social structures. Students from different regions have their traditions and they may believe in different religions. They have different opinions about today's political events, social behaviors, and economical development.

- **Difference in grading systems:** Each country has its own education structure; grading systems vary from country to country. The grading system we are using may not make sense in another country.

- **Difference in languages:** The language barrier is a major concern for globalizing an online course. Some of the languages do not even have corresponding translation for the new technology words.

- **Difference in financial systems:** The first thing that needs to be dealt with is the tuition. Before a program starts, there must be some answers to the questions such as how much to charge, how to set up an account, and how to exchange currencies.

- **Difference in time zones:** The time zone difference may cause difficulties in Web conferencing, student-instructor interaction, and collaborative learning activities. These difficulties can significantly limit the scale of globalization.

A globalized university is just like a globalized enterprise. We can learn a lot from a globalized enterprise. As the technology improves, some of the difficulties can be overcome now; for example, simultaneous translation from one language to another can be done electronically during a conversation. Most of the problems of globalization are not technical problems. They are management problems. The support of the administrators is critical to solving the problems such as the difference in grading systems and financial systems. One solution is to team up with universities in other regions to create a win-win management structure. In fact, many universities are doing it now. Some such programs fail, but many of them are successful. In the next few years, the trend will continue and more and more universities will work with other universities around the world to form globalized universities.

Learning Management System (LMS)

In the next few years, learning management systems will continue to improve. The following are some trends in the current LMS industry:

- To support globalization, a LMS will be constructed to allow language translation and allow users to upload the course content in their own languages.

- More and more templates or e-learning development suites will be developed for quick development of teaching materials and for class management.

- There will be more tools to support personalization by allowing students to manage their own course schedule and post information on their own blogs.

- For better support of collaboration, more tools will be added to LMSs. Tools for online live conversation and presentation will be more sophisticated.

- The future LMS software packages will have better protection by providing more security measures.

- There will be more reliable and better performance by supporting cluster computing. A cluster computing system will be built to support more students to simultaneously access and work on the system without slowing down the performance.

- More powerful databases will be used to support a LMS so that the LMS can be used as an information management center for a university.

- There will be more management tools included in a LMS so that the LMS can be used for student assessment and employee performance management, curriculum planning and scheduling, skill and knowledge management, and cost management.

- There may be some data analysis tools to identify behavior patterns and market trends for university management. These tools can also identify students' strengths and weaknesses, and identify employees who are best suitable for certain jobs.

The future LMS can do a lot more in the management of a university. It can cover the jobs in both Web-based teaching and human resources.

Online Collaboration

One of the major weaknesses of online teaching is the inconvenience for collaboration. To overcome the weakness, many researches have been done in this area and there have been many achievements. In addition to the currently available collaboration tools such as whiteboards, bulletin boards, chat rooms, and online presentation tools, the new tools such as collaboration effectiveness platforms, collaborative learning portals, and intelligent tutor systems are also under development.

- **Collaboration effectiveness platform:** This kind of platform can significantly improve interactivities among individuals, groups, and campuses. Within the platform, multiple communication media will be consolidated. Communication through different technologies such as Voice over Internet Protocol (VoIP), e-mail, audio and Web conferencing, and voice mail can be brought together and managed with unified messaging techniques.

- **Collaborative learning portal:** A collaboration learning portal is a Web site that combines application software, server software, Web services, communication services, and a search engine. It can combine different computing resources into an integrated information management and messaging platform.

- **Intelligent tutor system (ITS):** This is a system that mimics the way in which a real life tutor helps students. ITS can decide the proper exercise to be assigned to a student based on his/her understanding of the content. It can customize the tutoring style and monitor students' activities. Based on the answers given by a student, it will analyze why the student makes certain mistakes and come up with better ways to deliver the knowledge to the student.

Although online collaboration still has a long way to go to be adequate for online teaching and learning, a lot of successful stories have been reported about using these tools. Also, the younger generation students are more familiar with the online communication tools. To them, the current collaboration tools are not difficult. They like to use e-mail, instant messaging, and Web camps to communicate with each other. These habits make online communication much easier.

Development of Course Content

For developing online teaching materials, course content modularization has been considered as one of the trends that may have significant impact on Web-based teaching. The modularization process divides course content into small units. Then, based on the requirements of a course, it reassembles these small units to meet the course objectives. Through modularization, we can quickly develop new teaching materials as soon as there is a training request for a new technology. Modularization is particularly useful for the multidisciplinary courses which integrate multiple units from different topics. The integration can even be extended to include the units posted on different Web sites, created with different programming languages, and managed by different learning management systems.

Trends in Technology

Technology is critical for online computer lab construction. There are many technologies that are under development and will have a great impact on the development of online computer labs. In the following, let us have a closer look at these technologies.

Network Technology

As pointed out before, an online computer lab is a network-based computing system. The improvement of an online computer lab will depend on the improvement of the network technology. Several network protocols and devices will bring network computing to a new level. The following are some of them:

- **Ipv6:** An important network technology is Ipv6, which is the next generation protocol used by the Internet. The current version of the Internet protocol Ipv4 is about to run out of IP addresses. Ipv6 provides a large number of IP addresses, enough for the whole world to use for at least many decades. Ipv6 allows users to establish a high-quality path to transmit audio and video data required by Web-based teaching which needs high-quality transmission and fast performance. Interested readers can get more information about Ipv6 from Ipv6's (2006) Web site.

- **VoIP:** Another important protocol for computer lab construction is Voice over Internet Protocol and its devices. VoIP can greatly improve the quality of collaboration and reduce the cost of running an online course. VoIP Lowdown (2006) describes the use of VoIP in education.

- **WiMAX:** This is another technology that can make Web-based teaching and learning more effective and flexible. WiMAX is a wireless technology. Unlike the current wireless technology, WiMAX provides high-throughput broadband wireless communication over long distance. It can make mobile distance learning more flexible and affordable. More information about WiMAX can be found on Intel's (2006) Web site.

- **Personal broadcasting:** This technology borrows the idea from blogging. More than blogging, personal broadcasting provides live video Weblogging. The current trend in personal broadcasting is podcasting. The iPod has become a learning tool that receives podcasting. McCarthy (2006) mentions that universities such as Stanford University put the instructors' lectures on iTunes. Podcasting has been used to post short lecture notes at some universities. All the freshmen students at Duke University got free iPods provided by the university (McCarthy, 2006).

Some of these technologies are already available and others are still under development. More and more teaching and learning tools will be developed around these technologies. Once these technologies become widely available, Web-based teaching and learning will benefit greatly from them.

Computer Technology

In addition to the network technologies, an online computer lab can also benefit from the improvement of hardware and software.

Hardware Trend

Several hardware technologies will certainly have some impact on online computer labs. Devices built with these technologies will make servers run faster, store more data, and be easily accessed. They will also make client computers able to store more data and support more multimedia functionalities. The following are some of the technologies that are beginning to gain popularity.

- In the next few years, PCs will be more powerful. CPUs will have higher and higher clock speeds and the dual-core and quad-core CPU will be used by most personal computers.

- One of the significant movements is the use of tablet computers in higher education institutions. It has been reported that tablet PCs are given to freshmen engineering students to use at Virginia Polytechnic Institute (McCarthy, 2006).

- In the near future, the less power-consuming flash memory may find its way to replace small hard drives for notebook computers. A hybrid hard drive which is a combination of flash memory and a traditional hard drive will be used for better performance.

- As the price continues to decrease, there will be more network storage devices used by various organizations.

- In a few years, the newly developed high-definition DVD which can store 20GB to 50GB data on a single disk may be widely available for computer systems and game machines.

We will also find many new inventions in video and audio technologies and data storage technologies. Every year, when you look around, there will be many new technologies that can make computer systems get better and better.

Software Trend

Like computer hardware, computer software is also an area undergoing fast development. In the next few years, software installed in an online computer will have some major changes. Many current software packages will be upgraded and there will be many new software packages. Some of the potential changes are listed below:

- **Operating system:** A new generation operating system will be more secure. It will provide better user management and computer management tools. It will have better performance by reducing the start up and shutdown time and the time to load application programs to memory. The power management utility will be improved. There will be more tools to support video and audio devices. Wireless network will get better support. There will be better tools for managing digital video files. The support for Web browsers will also be improved. Many of the new operating systems will continue to support 64-bit computing.

- **Database management system (DBMS):** As application software, a DBMS will continue to be improved in the next few years. More and more DBMSs will support XML-based queries such as XPath or XQuery. More of the database management tasks will be done with GUI tools. There will be more security measures built inside a DBMS to protect data. In addition to SQL, more object-oriented programming languages such as C#, VB.NET, C++, and Java will be supported by a DBMS. Some of the DBMS packages will continue to improve the built-in data analysis tools such as data mining and Online Analytical Processing.

- **Social computing:** Social software can be used by the user to communicate with others to form an online community. Commonly used social software includes e-mail, instant messaging, group discussion, or blog. This type of software has a great potential to be used for collaborative learning. More tools will be available to establish social networks among students and instructors who will be able talk and see each other through the Internet. These tools will make tutoring easier for both instructors and students. They will allow instructors to get feedback from students immediately.

- **Educational gaming:** Educational gaming is a learning tool used by precollege schools. Gradually, it is becoming a learning tool for higher education. Educational gaming software can be used for courses such as gaming, computer graphics, security, multimedia, simulation, artificial intelligence, and many other technology-based courses.

- **Augmented reality:** This is a kind of technology that integrates the real world motion with a virtual environment. Following a person's head movement, a virtual environment will be generated for the person to experience the environment. So far, this technology is not widely used in e-learning. As the technology gets more mature, it can be used to simulate an IT environment which can be used for troubleshooting training and other hands-on practice.

- **Context-aware computing:** This is an adaptive mobile computing method. It changes computing context according to the computing environment such as the location of the user, people around the user, the time, and the hardware devices in use. There are some applications of context-aware computing. For example, context-aware computing can be used in a cell phone. When a person comes into a meeting room, his/her cell phone will automatically refuse to ring for an incoming call and will take a message instead. Context-aware computing has also been applied in e-learning as illustrated by Schmidt and Winterhalter (2004).

Final Thoughts on Technology Trends

In the above, some of the emerging technologies are being used in Web-based teaching, and some have existed for years and continue to improve themselves. There are many other technologies that an online computer lab can use to improve online teaching and learning. When creating computer labs, be sure to keep the new technology trends in mind. Not all of the new trends will fit in our online teaching environments. The online lab construction team needs to conduct experiments on a new technology before making it available to instructors and students.

Web-based teaching and technology are being improved so rapidly; it is not easy to predict what will happen in the future. One thing is for sure; there will be more new technologies next year. With the improvement of these technologies, the quality and performance of an online computer lab will get better and better. IT administrators, technicians, instructors, and students have to keep updating themselves to master these new technologies. All this makes online computer lab development challenging and exciting.

Conclusion

In this chapter, the trends in Web-based teaching and technologies have been explored. This chapter first investigated the trends in the framework of Web-based teaching. It discussed the trends in the areas of e-learning structure, management, online collaboration, and content development. In each area, it investigated the trends and the possible use by an online course. It also discussed the significant impact of the trends on Web-based teaching. The second topic in this chapter is the trends in network technology and computer technology. Several new network trends such as Ipv6, VoIP, WiMAX, and personal broadcasting were discussed. For computer technology, the discussion was divided into the trends in hardware and the trends in software. For hardware, the trends in CPUs, Tablet PCs, storage devices and high-definition DVD were examined. For software, the trends in operating systems, database management systems, social computing, education gaming, augmented reality, and context-aware computing were discussed.

We are now entering a new era of teaching and learning. As mentioned in this chapter, Web-based teaching and technologies will continue to be improved. No matter what hardware and software have been installed in a computer lab, in a few years they will be out-of-date, and there is no sign that progress is slowing down. Therefore, all the people involved in the construction of online computer labs need to keep up with the trends.

References

Camacho, D., & Rodríguez-Moreno, M. D. (in press). Towards an automatic monitoring for higher education learning design. *International Journal of Metadata, Semantics and Ontologies*. Retrieved April 9, 2007, from http://www.ii.uam.es/~dcamacho/papers/dcamacho-IJMSO.pdf

Intel. (2006). WiMAX broadband wireless technology access. Retrieved April 9, 2007, from http://www.intel.com/netcomms/technologies/wimax/index.htm?ppc_cid=ggl|wng_wimax|kC74|s

Ipv6. (2006). Retrieved April 9, 2007, from http://www.ipv6.org

Koper, R. (2004). Use of the Semantic Web to solve some basic problems in education: Increase flexible, distributed lifelong learning, decrease teacher's workload. *Journal of Interactive Media in Education, 6*, 1-9.

McCarthy, C. (2006). Tablet PCs required for Virginia Tech engineers. Retrieved April 9, 2007, from http://news.com.com/Tablet+PCs+now+required+for+VA+Tech+engineers/2100-1041_3-6090046.html

New Media Consortium & EDUCAUSE Learning Initiative. (2006). The horizon report 2006 edition. Retrieved April 9, 2007, from http://www.nmc.org/pdf/2006_Horizon_Report.pdf

Paulsen, M. F. (2005). Online education: Trends and strategic recommendations. Retrieved April 9, 2007, from http://home.nettskolen.nki.no/~morten/pp/CampusInnovation.ppt#541,1

Ray, R. (2006). Top 2006 technology trends for small business. Retrieved April 9, 2007, from http://www.smallbiztrends.com/2006/01/top-2006-technology-trends-for-small-business.html

Reiser, R., & Dempsey, J. V. (2006). *Trends and issues in instructional design and technology* (2nd ed.). Upper Saddle River, NJ: Prentice Hall.

Schmidt, A., & Winterhalter, C. (2004). User context aware delivery of e-learning material: Approach and architecture. *Journal of Universal Computer Science, 10*(1), 28-36.

Sloan Consortium. (2005). Growing by degrees: Online education in the United States, 2005. Retrieved April 9, 2007, from http://www.sloan-c.org/publications/survey/survey05.asp

Ullrich, C. (2005). Course generation based on HTN planning. In *Proceedings of the 13th Annual Workshop of the SIG Adaptivity and User Modeling in Interactive Systems* (pp. 74-79).

VoIP Lowdown. (2006). VoIP in education. Retrieved April 9, 2007, from http://www.voiplowdown.com/2006/01/voip_in_educati.html

About the Author

Dr. Lee Chao is currently a professor of math and computer science in the School of Arts and Sciences at University of Houston-Victoria (USA). He received his PhD from the University of Wyoming (USA), and he is certified as an Oracle Certified Professional and Microsoft Solution Developer. His current research interests are data analysis and technology-based teaching. In addition, Dr. Chao is also the author of over a dozen research articles and a book on database development and management.

Index

X